Poverty Alleviation and Poverty of Aid

Aid effectiveness has emerged as an intensely debated issue amongst policy makers, donors, development practitioners, civil society and academics during the past decade. This debate revolves around one important question: does official development assistance complement, duplicate or disregard the local resource endowment in offering support to recipient economies?

This book draws on Pakistan's experience in responding to this question with a diverse range of examples. It focuses on a central idea: no aid effectiveness without an effective receiving mechanism. Pakistan is among the top aid recipient countries in the developing economies. It was a shining model in the sixties and it ranks among the highly underperforming countries after the new millennium. This book offers an insight into the dynamics of success and failure of Pakistan in availing foreign financial and technical assistance for human development and poverty alleviation. It draws on field experiences to present case studies on water, shelter, health, education, and health and safety at work to identify the causes and consequences of aid in relation to social reality. Findings relate to developing economies and would be of interest to a wide range of individuals within the development sector.

Fayyaz Baqir has served as CEO of the Trust for Voluntary Organizations (TVO), Senior Civil Society Advisor of the United Nations, and a development professional at Aga Khan Rural Support Programme (AKRSP). He has researched and taught at McGill University, Georgetown University, Harvard University, University of Idaho, Wellesley College, Tilburg University (The Netherlands), Gothenburg University (Sweden), Quaid e Azam University, Punjab University and National Defence University (Pakistan) on themes relating to inclusive governance, participatory development and sustainable change. He received a Top Contributors' Award from UNDP's Global Poverty Reduction Network in 2007 and 2008 and an Outstanding Performance Award from UNDP for creating a vibrant small grants programme for low-income communities in Pakistan. He has travelled to more than 40 countries in Asia, Africa, Europe, and North and South America as part of his professional work.

Routledge Studies in Development and Society

For more information about this series, please visit: www.routledge.com/Routledge-Studies-in-Development-and-Society/book-series/SE0317

Poverty Alleviation and Poverty of Aid

Pakistan

Fayyaz Baqir

Routledge
Taylor & Francis Group

LONDON AND NEW YORK

First published 2019 by Routledge
2 Park Square, Milton Park, Abingdon, Oxon OX14 4RN

52 Vanderbilt Avenue, New York, NY 10017

First issued in paperback 2020

Routledge is an imprint of the Taylor & Francis Group, an informa business

British Library Cataloguing-in-Publication Data
A catalogue record for this book is available from the British Library

Library of Congress Cataloging-in-Publication Data
A catalog record has been requested for this book

ISBN 13: 978-0-367-58782-6 (pbk)
ISBN 13: 978-1-138-48098-8 (hbk)

Typeset in Times New Roman
by Apex CoVantage, LLC

Dedicated to

A Stranger in the New World

Contents

Illustrations

Figures

Tables

Boxes

Preface

This book is an attempt to reveal the tacit knowledge that I harvested over the past 30 years travelling every nook and corner of Pakistan as a development professional. It gave me the opportunity to unlearn many things I had learned during my schooling and learn many new things known only to the toiling, creative and illiterate people of my country. I am sharing it in the hope that it will ignite the spark in many caring hearts who believe that "anyone can make the difference" and it begins with the self. It is also a tribute to some of my teachers who taught me how to feel, think and act. The summary of all that I learned during my years in the field is that for poverty alleviation you don't need a big purse or big mind but a big heart.

Fayyaz Baqir
McGill University
February 2017

Acknowledgements

I want to thank the O'Brien Fellows in Residence programme for providing me the opportunity to write these case studies reflecting my three decades of work among low-income communities in Pakistan in different capacities. This research would not have been possible without the help of many individuals and institutions. I want to express my appreciation to the Centre for Human Rights and Legal Pluralism (CHRLP) and Institute for Studies in International Development (ISID) of McGill University for hosting my stay for one year and providing me the opportunity to exchange views on my draft chapters with the students and the faculty. Special and sincere thanks are due to Professor Nandini Ramanujam, Professor Francois Crepeau and Professor Sonia Laszlo, whose continuous encouragement kept me on track during the course of my work until its completion.

Professor Madhav Badami provided very incisive comments for improving my case study on water. I am heartily grateful to Ms. Sharon Webb and Sheryll Ramasahi for their constant help during my stay at McGill Faculty of Law, and all other colleagues for their kind assistance and encouragement during the research. Parisa Akberimalkeshi helped me during the last stages of my work in dealing with the formatting requirements. Special thanks are due to my wife Rehana Hashmi for her continuous support and encouragement.

Fayyaz Baqir

Abbreviations

ABES	Adult Basic Education Society
ADB	Asian Development Bank
AKDN	Aga Khan Development Network
AKRSP	Aga Khan Rural Support Programme
APC	Agricultural Prices Commission
APPNA	Association of Pakistani Physicians in North America
APTMA	All Pakistan Textile Mills Owners Association
BHUs	Basic Health Units
CBO	Community-Based Organization
CDGK	City District Government Karachi
CDPI	Centre for Peace and Development Initiatives
CEDAW	Convention on Elimination of All Forms of Discrimination Against Women
CO	Community Organization
CRC	Convention on the Rights of the Child
CRI	Cotton Research Institute
CRP	Community Resource Person
CSR	Corporate Social Responsibility
DA	Development Administration
DCG	District Core Group
DFID	Department for International Development
EPB	Export Promotion Bureau
FFS	Farmer Field School
FSC	Federal Shariah Court
FSCD	Federal Seed Certification Department
GDP	Gross Domestic Product
GSP	Generalized Scheme of Preference
GST	General Sales Tax
HDA	Hyderabad Development Authority
HANDS	Health and Nutrition Development Society
HBFC	House Building Finance Corporations
HUD&PHED	Housing, Urban Development and Public Health Engineering Department

ICCPR	International Covenant on Civil and Political Rights
ICESCR	International Covenant on Economic, Social and Cultural Rights
IFIs	International Financial Institutions
ILO	International Labour Organization
IMF	International Monetary Fund
IPM	Integrated Pest Management
JBIC	Japanese Bank for International Cooperation
JETRO	Japan External Trade Organization
JICA	Japan International Cooperation Agency
KA	Katchi Abadis
KAWWS	Karachi Administration Women Welfare Society
KDA	Karachi Development Authority
KMC	Karachi Municipal Corporation
KWSB	Karachi Water and Sewerage Board
LHW	Lady Health Worker
LIFE	Local Initiative Facility for Improvement of Urban Environment
MDG	Millennium Development Goal
MTDF	Medium Term Development Framework
NATO	North Atlantic Treaty Organization
NCHD	National Commission for Human Development
NGO	Non-Government Organization
NRSP	National Rural Support Programme
ODA	Official Development Assistance
OPP	Orangi Pilot Project
PAPA	Pakistan Agriculture Pesticide Association
PARC	Pakistan Agricultural Research Council
PARD	Pakistan Academy for Rural Development
PCCC	Pakistan Central Cotton Committee
PIPHCMP	Punjab Integrated Primary Healthcare Model Programme
PIU	Program Implementation Unit
PMU	Programme Management Unit
PPAF	Pakistan Poverty Alleviation Fund
PPHIC	President's Primary Healthcare Initiative
PRSP	Punjab Rural Support Programme
PTA	Parents Teachers Association
RHC	Rural Health Centre
BHU	Basic Health Unit
RSP	Rural Support Programme
SAP	Social Action Programme
SCPEB	Society for Community Support for Primary Education in Balochistan
SDG	Sustainable Development Goal
SFV	Specialized Financing Vehicle
SGA	Sindh Graduates Association
SHG	Self-Help Group
SIUT	Sindh Institute of Urology and Transplantation

SKAA	Sindh Katchi Abadis Authority
SLGO	Sindh Local Government Ordinance
SMC	School Management Committee
SOP	Standard Operating Procedure
SPDC	Social Policy and Development Centre
SSK	Shoaib Sultan Khan
SUCCESS	Sindh Union Council and Community Economic Strengthen Support
TBA	Traditional Birth Attendant
TEP	Teacher's Education Programme
TMA	Tehsil Municipal Administration
TTDC	Thana Training and Development Centre
TVE	Technical and Vocational Education
UCBPRP	Union Council Based Poverty Reduction Programme
UNDP	United Nations Development Programme
URC	Urban Resource Centre
Village AID	Village Agricultural and Industrial Development
VIM	Village Improvement Model
WSP	Water and Sanitation Programme

1 NGO's ladder to development – knowledge and the path of solicitation

How could the rich world possibly take responsibility for billions of people outside their borders . . . Happily, they have reasonable answers . . . they can be achieved within the limits that the world has already committed: 0.7 percent of the gross national product of the high-income world.

—Jeffrey Sachs (2005: 288)

If the United States were to hike its foreign aid budget to the level recommended by the United Nations – 0.7 percent of national income – it would take the richest country on the earth more than 150 years to transfer to the world's poor resources equal to those they already possess.

—Hernando De Soto (2000: 5)

Pakistan received $58 billion in foreign aid from 1950–99; however, it systematically under-performed on most of the social and political indicators. Pakistan was the third largest recipient of ODA after India and Egypt during 1960–98. If it had invested all the ODA during this period at a real rate of 6 percent it would have a stock of assets equal to $239 billion in 1998, many times the current external debt.

—Williams Easterly (2001)

If you are not confused, you are not in UNDP.

—UN Resident Representative Pakistan

Want to help someone? Shut up and listen.

—Ernesto Sirolli (2012: 1)

Development ideology

The role of external economic assistance seems to be very confusing in view of the conflicting views on virtues and vices of foreign aid cited in the chapter epigraph. Effectiveness of external assistance can be well understood if we take into consideration the complete context of aid for developing economies after the Second World War. The end of the war triggered the search for identity in the former colonies. Colonies needed to discover the connection between their newly acquired freedom and development of their economies. They needed to define their development goals and select the best available means to achieve them. One of the key questions at hand was to determine what choices were to be made during the course of accepting and eliminating the difference in following the path of advanced Western economies (Rapley: 2007). In South Asia, the Americans filled the void created by the departure of British colonial rulers. Americans, like their predecessors, thought developing native societies entailed reproducing them in their own image. To American advisors, assisting the developing world meant helping them in copying the American style of life, promoting consumerism and industrialization, proposing policies based on free market solutions, making generous investments through private sector and moulding political, academic and social institutions in the American way. This ideology of development copying did not have much success in the case of Pakistan. It led to mass upheavals and explosive conflicts at the end of first decade of American-assisted development.

The extremely limited capacity of Pakistan's political class in conducting politics in conformity with the democratic norms led it to strengthen the civil and military bureaucracies in defining and implementing the development agenda. Contrary to Hamza Alavi's observation that Pakistan inherited an "overdeveloped state" (Alavi: 1972), Pakistan's ruling class consisting mainly of the landed elite 'overdeveloped' the state itself to compensate for its own deficiencies in following the democratic discourse (Khan: 1967). The state institutions inherited and developed by Pakistan consisted of a strong law and order arm and lacked a development administration. According to American advisors, all these problems could be fixed by pumping in development aid. The problem was that Pakistan did not have a "receiving mechanism" in place to convert American dollars to the American way of life in Pakistan (Khan: 1980). A national security state that emerged after freedom could not follow the norms of a modern nation state or a welfare state. Strengthening defence capability and creation of inequalities for capital formation were two key ingredients of Pakistan's "development agenda" from day one (Haq: 1966).

Since no capitalist class existed in Pakistan to implement a capitalist development agenda, American economic assistance was channelled through the state to create a capitalist class. The state-led capitalism found favour with Pakistan's American advisors because in the early post–Second World War period 'statism' under the influence of Keynesian thought emerged as a leading ideological tendency in development policy and practices (Rapley: 2007: 28). Due to limited success of this tendency, two decades later structural adjustments were tried as a panacea to the ills caused by slow growth and increasing inequalities in the post-colonial economies. Both approaches had extremely limited impact due to the mismatch between the American and Pakistani institutional landscapes. There have been two responses to these failures. The first, according to the conventional American wisdom, is to ask Pakistan to "do more" of the same. The second response, deliberated upon and tested by development practitioners, has been to see where the copying approach goes wrong. It is important to note here that the similarities between American and Pakistani state, society and social institutions are very superficial. A review of dissimilarities provides better insight in dealing with the problem of development and poverty alleviation than the similarities. This critical dimension has been neglected by external advisors and has had an adverse impact on the efforts made to address these issues.

Looking at the dissimilarities, we note that Pakistan at the eve of independence did not have a business class and Pakistan's industrial base consisted of one small factory. Political leadership had extremely limited experience of public service except for a small group around Pakistan's founding father, Mohammad Ali Jinnah. Civil and military institutions had to be built almost from scratch. The middle class, mostly consisting of Hindus and Sikhs, had migrated to India. Pluralism, secular outlook and openness of early years did not last long. Pakistan's first national anthem, written on the desire of Jinnah by Jaggan Nath Azad, a Hindu of the Saraiky ethnic group, was replaced by a new anthem soon after his death. The tradition of multi-faith leadership also did not last long. It is important to note here that while the founder of the All India Muslim League was Ismaili Imam Sir Sultan Mohammad Shah Aga Khan, the creator of the idea of Pakistan, Mohammad Iqbal, was a Sunni; the first governor general was a Twelver Shia, Mohammad Ali Jinnah; and the first foreign minister was an Ahmadi, Sir Zafarullah Khan. This pluralism did not last long, and minorities gradually lost the protection, respect and tolerance that they received at the eve of independence. It depicted a serious leadership challenge in the course of Pakistan's journey to development and prosperity.

Historical context of development challenges

Pakistan was haunted by the existential dilemma of being or not being at peace with its neighbour at the eve of independence. The people of Pakistan were condemned to reap the harvest of fear in the season of hope. The leaders of the Muslim League, the Indian National Congress and the British rulers had all agreed to partition India in a cordial and peaceful manner supposed to guarantee good neighbour relations between the two independent secular states succeeding British India. However, the partition was carried out in a hurry, one year earlier than planned. Boundaries were not marked as agreed in principle, proper security arrangements were not made for mass movement of population across the new international borders, and the masses were not prepared to fulfil their responsibilities as citizens of newly independent states (Ali: 1967). It led to unprecedented bloodshed, mass killings, rape, looting and plunder. Communal frenzy reached such a level that Mahatma Gandhi, who had declared to fast to death to prevent clashes between Hindus and Muslims, was killed at the hands of a Hindu fundamentalist. Soon after the partition, India blocked water flow from every canal in India flowing to Pakistani Punjab to strike terror in the hearts of millions of Pakistani farmers depending on agriculture for their survival (Jamal: 2017). These circumstances, combined with the death of Pakistan's founding father, Moham-mad Ali Jinnah, soon after independence and political inexperience of his succes-sors, led to generate a development vision inspired by the idea of national security.

The key challenge faced by Pakistan was to find ways to respond to opportuni-ties for catering to unmet development needs. The reality was that there was no 'receiving mechanism' in place to receive and use development funds. As aptly pointed out by Pakistani development practitioner Dr. Akhter Hameed Khan (1914–1999), three key infrastructures – administrative, social and political – were required to create a receiving mechanism for developing the economy to alleviate poverty, generate income opportunities and promote growth. Pakistan had a cash-starved economy due to the existence of a wide barter economy in the agriculture sector; consequently, it had a narrow tax base and very limited fiscal space for carrying out the 'development project'. The entrepreneurial class was also non-existent. A large number of merchants who migrated from India had no experience of running industrial enterprises. Big landowners thrived due to own-ership of vast tracts of land despite low farm productivity and absentee farm man-agement. That is why it suited Pakistan to follow the path of state-led capitalism. The architect of Pakistan's economic planning, Dr. Mahbubul Haq (1934–1998), pointed out that Pakistan had to fill two gaps to take off as a modern industrial economy: a saving gap and a foreign exchange gap. Saving needed to finance investments was to be generated by creating income inequalities through a fiscal regime based on subsidies, price and foreign exchange control and import restric-tions. These inequalities sounded unfair, but in the long run they would alleviate poverty by generating greater employment opportunities and increasing wages due to high demand for labour. Foreign exchange needed for purchase of capital equipment and industrial inputs was to be acquired through external economic assistance, control of the exchange rate and borrowing. In view of this policy per-spective, the public sector and Planning Commission played a very important role in promoting the private sector and "free market economy" through strict market control in Pakistan (Baqir: 1984).

The development strategy followed by Pakistani decision makers was to achieve economic goals through political means under the façade of the market (Baqir: 1984). Public policy was the key instrument of growth in Pakistan. The policy was top down and discriminatory and led to strengthening of a patron cli-ent culture. The lack of a receiving mechanism combined with lack of knowledge of local conditions and an uneven playing field led to policy failure in fulfilling the "American dream" in Pakistan. The free market approach was beset by two challenges: restricting internal and external competition. Competition with the outside world was eliminated to protect the nascent business class against expe-rienced foreign businesses. It was carried out in the name of import substitution through an elaborate tariff and quota regime. Internal competition against local small producers was restricted through licencing and price control. It resulted in substitution of industrial goods for the goods produced by hundreds of thousands of local artisans and small producers and wiping out the artisan class. It led to the creation of a parasitic capitalist class, not a vibrant and competitive bourgeoisie.

Policy regime for 'capitalist' development

National security state, aid and income inequalities

Pakistani planners had two objectives in mind: creating a strong defence and a strong industrial base through international borrowing, and creation of domestic income inequalities. This concept of development hinged on the development of a national security state. Pakistan inherited a resource base comprising scarce capital resources and relatively abundant human resources, but the development policy did not take into consideration this resource endowment in planning for development. Pakistan's early planners neglected human resource development and designed a development strategy based on capital accumulation through transfer of resources from agriculture, small-scale production and less developed regions to industry and urban centres. This was an exclusionist strategy. The real burden of this transfer of resources was borne by the agricultural and industrial working class. This transfer of resources led to rapid development of physical capital at the neglect of human and social capital. During the first three decades Pakistan's agricultural and manufacturing sectors underwent a rapid transformation with creation and expansion of large-scale industry, the use of high-yielding varieties of seeds and fertilizer as well as mechanization in agriculture. This process was simultaneously a process of capital accumulation, structural transformation and growth of output. It had two fatal flaws. First, the business class leading the process of industrialization was extremely protected and added negative value in the process of production and could not help Pakistan fill its foreign exchange gap through export of consumer goods in the international market. Second, the inequalities created in the process of transfer of resources to industry created deep-seated resentment between the wage and salary earning classes, the rural population and inhabitants of less developed regions on the one hand and the emerging industrial bourgeoisie on the other (Baqir: 1984).

Protection of the 'infant industry' included, among other things, overvaluation of Pakistani currency, tariff and quota restrictions on import of consumer and capital goods, licencing for the import of raw materials and equipment, and provision of subsidies on export of industrial commodities with the net impact of transfer of resources from agriculture to industry, and from less developed regions to more developed regions (Islam: 1981; Lewis: 1969, 1970). Mahbub ul Haq, deputy chairman of the Planning Commission during Ayub Khan's regime, persuasively argued the case for generating inequalities to fill the saving gap to promote growth in Pakistan and external borrowing for filling the foreign exchange gap. In his view, foreign economic assistance in the absence of sizable domestic savings provided the opportunity for 'development without tears'; and striving for equal distribution of income amounted to aiming for 'equal distribution of poverty', so an increase in income at the expense of efficiency and equity was the best choice available to Pakistan. The idea of the saving capacity of the poor was not known to the mainstream economists at that time. Decades later, Hernando de Soto pointed out that "The value of savings among the world poor is, in fact, immense – forty times all the foreign aid received throughout the world since 1945" (De Soto: 2000: 5). The poor not only have enormous saving capacity; their remittances from abroad in recent years have also filled Pakistan's foreign exchange gap, the role which was supposedly to be played by the business class called the 'robber barons' by Gustav Papanek (Papanek: 1967).

Ironically, Haq was the first one to point out the concentration of industrial assets in the hands of 22 families in Pakistan. Mass movement against the authoritarian and exploitative rule of General Ayub Khan was built on the slogan of ending the monopoly of the 22 families. These exclusionist policies and practices continued in the form of exchange rate control in 1960s; nationalization of small production and service providing enterprises in 1970s; expansion of the black economy due to massive arms and drugs trade during the "free world's" war against Soviet invasion in Afghanistan in the 1980s; and "privatization" of education, health, security and other basis services due to the state's neglect of basic needs of the poor and expansion of informal sector in the 1990s and 2000s. These inequalities did not result in self-sustained growth as envisioned. However, the American advisors and Pakistan's policy makers were aware of the need to address the question of mass poverty linked with the pattern of development unfolding in Pakistan. They tried to overcome poverty by increasing farm productivity and provided assistance under the Village AID programme for this purpose. While the Planning Commission followed the policy of filling the financial gaps for economic development, the Village AID programme in East Pakistan conducted an experiment to fill the institutional gap to alleviate rural poverty. Interplay between these two policy regimes had a far-reaching impact on political and economic stability in Pakistan.

Institutional landscape

State as agent of modernization

Both the processes of economic development and poverty alleviation were to be led by the state. While leading American experts saw the state as the agent of modernization, critic of the capitalist path of development Hamza Alavi presented the idea of an overdeveloped state playing relatively autonomous economic role by offsetting the influence and power of each section of the ruling class. Alavi argued that Pakistan's pattern of development could be understood through the prism of an overdeveloped state (Alavi: 1972). Both these views failed to look at the soft belly of the state in implementing the development project. Contrary to Alavi's perception of inheriting a 'metropolitan' structure, Pakistan inherited a truncated army and civil service (Khan: 1967) and a single factory to represent the 'power' of the indigenous bourgeoisie. American experts, on the other hand, ignored the fact that there was no development administration in existence in Pakistan. In place of a development administration, the British Raj depended heavily on local landed elite. It was the case in China and other British colonies in Asia as well (Tung: 1926). Most of the administrative work relating to construction, repair and maintenance of public works, dispensation of justice and maintenance of law and order is carried out by the landed elite to date across many communities in the Asia Pacific region (Khan: 1996; Golub: 2003; World Bank: 2008; Faure: 2011).

Limitation of state-led capitalism financed through external borrowing arises because the state as a modernizing agent has serious capacity gaps. According to Akhter Hameed Khan, the Pakistani state is a state with an extremely weak coordination mechanism for delivery of services to the citizens. It depicts managerial incompetence. De Soto has noted similar limitations in the cases of the Philippines, Egypt and Peru (De Soto: 2000). The Pakistani state lacks the capacity to generate surplus (Haq: 1966). Pakistan has a much smaller size of its social sector budget compared to the size of indigenous philanthropy, and is considered a 'failed' state due to the lack of capacity to generate and collect taxes to meet its revenue requirements (Martinez-Vazquez and Cyan: 2015).

William Easterly pointed out the limitation of the Pakistani state even in making judicious use of economic resources received for economic development. In Easterly's words,

> Pakistan received $58 billion in foreign aid from 1950–99 but systematically under-performed on most of the social and political indicators. Pakistan was the third largest recipient of Official Development Assistance (ODA) after India and Egypt during 1960–98. If it had invested all the ODA during this period at a real rate of 6 percent it would have a stock of assets equal to $239 billion in 1998, many times the current external debt.
>
> (Easterly: 2001)

The Pakistani state also has extremely limited capacity to raise taxes to follow a path of sustained growth. As noted by Cyan,

> The tax system of Pakistan continues to under-perform particularly in its ability to raise adequate revenues. The bases of the most important taxes, such as Personal and Corporate Income Tax and the General Sales Tax (GST), continue to be narrow and the level of tax evasion remains high. Moreover, in recent years, the Tax-to-GDP (Gross Domestic Product) ratio has seen a substantial decline.
>
> (Martinez-Vazquez and Cyan: 2015)

Naviwala has mentioned the IMF's assessment that "for countries to finance the Millennium Development Goals, they must raise 20 percent of their national income through taxes" (Naviwala: 2016: 9). It is important to note here that Pakistan only collects 9 percent of its GDP as taxes.

Gunnar Myrdal was of the view that Pakistani state was a soft state because it inherited many liabilities at the time of partition (Myrdal: 1968: 311). Because a weak political culture undermines the power of the state, it does not lend it greater autonomy. The leadership of the Muslim League found little time for political preparation in the years leading to the creation of the new state of Pakistan (Myrdal: 1968: 325). In the words of the Muslim League's founding president, Sir Sultan Mohammad Shah Aga Khan, when Jinnah created Pakistan, he had no programme for the new nation. He had only one programme: "follow me."[1] According to Myrdal, a military coup in Pakistan was not an attempt to redistribute power but a return to the norms of decency and public service and regrouping of the old ruling class. From this point of view, it makes more sense to look at military takeover as an indication of the softness of the state rather than a sign of an overdeveloped structure, not only in Pakistan but also in many other Asian economies. This background explains the under-achievement of state-led capitalism in Pakistan and highlights the need to align economic assistance with institutional realities and practices.

Three essential infrastructures

Pakistan's renowned development practitioner Akhter Hameed Khan identified three essential infrastructures – administrative, political and socioeconomic – for poverty alleviation and local development. He noted that Pakistan had inherited a good law and order administration, but it did not have any development administration in place below the district level to guide and support majority of population in rural areas. Police stations – the lowest tier of law and order administration – had been situated in such a way that they had reach out capacity to every village in the empire. It was essential to devolve the so-called nation building or line departments like the police or revenue departments. During the British Raj, due to the absence of line departments below the district level, big landlords or tribal chiefs carried out most of the work relating to development, repair and maintenance of local infrastructure. This created the institutional base for interdependence of the state and civil society. Soon after independence, the feudal structure in East Pakistan started crumbling due to outmigration of Hindu landowners to West Bengal. This led to disrepair and lack of maintenance of water channels and public works in rural areas. No social infrastructure to fill this institutional vacuum existed. This led to underperformance of the rural economy and mass poverty and pointed to the need for creating a social infrastructure to work in collaboration with the administrative infrastructure (Khan: 1996). Political infrastructure was needed to provide assistance for accessing resources for fulfilling the expressed needs of rural communities identified through the social infrastructure. State-led capitalism could not deliver on the ground as these essential infrastructures constituting the 'receiving mechanism' were missing when the development project started.

Development administration

Advisors from the Ford Foundation and Harvard were aware of the serious institutional gaps undermining Pakistan's capacity for economic development and poverty alleviation. To address this issue, the Village AID programme was launched in Pakistan.[2] While the political and administrative leadership in West Pakistan ignored the institutional gaps and used the development funds as handouts, Dr. Akhter Hameed Khan, who was picked to lead this programme in East Pakistan, endeavoured to build the administrative and social infrastructure to create an appropriate receiving mechanism for self-sustained rural development and poverty alleviation. In West Pakistan, even the British legacy of the law and order administration was not maintained. What distinguished British officers from their Pakistani successors was their practice of regularly visiting the field and meticulously documenting their findings in the form of maps, gazetteers, diaries and field notes. Discontinuity of this practice along with rapid expansion of urban population led to a scanty knowledge base on local communities in urban areas as well.

The first requirement for a development administration (DA) is to create a viable unit of administration. This unit, like the police station, should reach out to all the surrounding villages and serve as a mechanism to build coordination between all the relevant government departments. Dr. Khan decided to address this concern through the Village AID (Village Agricultural and Industrial Development) programme in 1959 as director of the Pakistan Academy for Rural Development (PARD) situated at Commilla. He led the process of devolving and coordinating the nation-building department to the *thana* (police station) level by establishing the Thana Training and Development Centre (TTDC). The programme was supported by the Harvard Group, Michigan State University, the Ford Foundation, the Planning Commission and the World Bank. The programme spread to 50 percent thanas (lowest unit of administration with direct outreach to villages) of former East Pakistan and was internationally acknowledged for its outstanding performance. Within three years, 417 thanas in East Pakistan had training and development centres. PARD Commilla experience was repeated by PARD Peshawar in Daudzai Thana of the North Western Frontier Province (NWFP) in 1972 and produced impressive results (Khan: 1980).

Dr. Khan demonstrated through TTDC that the district as well as the *tehsil* (or subdivision or *taluka*) was not a viable unit for DA. It had to be the thana. DA envisages direct access of services and supplies to the people. District and subdivision, tehsil or taluka headquarters are too far removed from the village or the *mandi* (market) towns now emerging through urbanization. The major ideas incorporated in establishing TTDC were to revitalize district administration, create democratically constituted local self-government, divide business and administration, co-coordinate government departments adequately both horizontally and vertically and create task organizations for development at local level.

Between 1959 and 1971, Dr. Khan demonstrated TTDC's success in 50 percent of thanas in former East Pakistan, but the perception of inequalities caused by statist development policies outweighed the sense of wellbeing caused by the massive achievements of TTDCs. As a result, East Pakistan separated from West Pakistan. Dr. Khan moved to West Pakistan (current Pakistan), and his first experiment in building an administrative infrastructure for rural development in (West) Pakistan was carried out in 1972 in Daudzai. Shoaib Sultan Khan, director of PARD Peshawar, hosted the programme. PARD realized that the existing jurisdiction of the police station could be converted into a viable unit for administration, being viable for (1) provision of services and (2) upgradation of skills. The academy set out to persuade the line departments to expand to Markaz (to be established at the police station level).[3] This experiment was short-lived and ended with the exit of Shoaib Sultan Khan from service due to a vilification campaign launched against him.

Shoaib Sultan Khan followed this approach when he led AKDN's rural development initiative Aga Khan Rural Support Programme (AKRSP) in Gilgit district in 1982. AKRSP was a civil society version of a development administration and termed a support organization. Eventually this approach expanded through establishment of Rural Support Programmes (RSPs) all over Pakistan.[4] Federal and provincial governments provided endowment funds for all these initiatives. However, this apparent success should not be misinterpreted. Even with a very wide outreach, RSPs cover between 5 percent and 10 percent of the rural population. Also, this success has not led to civil service reforms, a task which was undertaken by PARD Comilla. Pakistan is still in need of something like PARD Comilla to create a suitable receiving mechanism to make effective use of domestic and external official resources for poverty alleviation. The donor's shortcut invention to deal with this challenge is to create expensive Programme Management Units (PMUs) for their projects which have had a questionable record in achieving targets.

Social infrastructure

Social infrastructure plays an important role in helping poor households access services from the administrative infrastructure. Organized communities significantly reduce the demand on the development administration's resources by creating economies of scale in accessing and utilizing technical guidance from thinly spread government departments. Such communities also take over the work previously done by the contractors and do it under a much lower budget. Known generally as civil society, social infrastructure can be broadly divided in two categories: community organizations and support organizations. Community organizations constitute the lowest rung of the ladder. Their essential role is to organize beneficiaries, improve skills of local professionals and encourage savings to raise capital. Support organizations provide technical and social guidance and credit to community organizations so that they can realize their potential for development.

At present, there are more than 65,000 registered and more than 100,000 unregistered community organizations in Pakistan. Such a large-scale presence of CSOs did not exist in 1947. In fact, British policies led to eradication of a well-developed social infrastructure in place in the 1850s. During the British Raj, communities were disenfranchised due to (1) elimination of community trusts, (2) replacement of the elected tribal chief's authority with nominated and hereditary chiefs and (3) establishment of land titles by converting the commons into agricultural land holdings through establishment of canal colonies, and through conferral of land titles to big landowners in the land settlement process. Social infrastructure has developed in Pakistan with the passage of time, but it still reaches out to a small fraction of Pakistan's population.

Support organizations become the key links between government and communities for service delivery and use their innovative models for advocacy. However, there are many distinct categories of CSOs. Adil Najam has pointed out that there are four distinct subspaces in the wider space of interaction between the state and civil society: cooperation, confrontation, complementarity and co-optation. Their ends and means govern the relationship between them and the state. They may have (1) similar goals and similar means, which would lead to a cooperative relationship; (2) similar goals and dissimilar means, resulting in collaboration; (3) dissimilar goals and similar means, in which case engagement is possible only through co-optation; and (4) dissimilar goals and dissimilar means, where claiming the rights would lead to confrontation (Najam: 2000).

The extent to which CSOs succeed in seeking access to state and market resources is not hampered due to power or resource deficit of CSOs but due to their undeveloped capacity to overcome trust deficit. In many cases, we see that the fiscal and legal space conceded by the state to the communities is not appropriated by the civil society, and communities make rent payment to 'uncivil society' to appropriate this space through extra-judicial means. The question is, what explains this disconnect between the state and the civil society? As aptly stated by Tony Beck, "search for solutions by these CSOs flow like the mythical river in *Mahabharata* upwards" (Beck: 1994: 115). Uncivil society succeeds when it creates solutions in line with the ground reality, and this is a necessary condition for the success of CSOs as well.

Political infrastructure

The political infrastructure comprises all the elected bodies from the local to the federal government. Local government constitutes the most important tier of political infrastructure for creating community access to public resources. Unfortunately, local government has had a chequered life in Pakistan. It was always introduced by military governments to seek outreach and legitimacy at the grassroots level and rolled back as soon as elected governments took power. Successive devolution plans in Pakistan mainly concentrated on the local government and tried to empower it by transferring many of the powers from higher tiers of government. This transfer of power took place in a half-hearted fashion and did not ensure access of services and supplies to the grassroots. The most conspicuous element missing from these arrangements was devolution of fiscal authority. In the absence of fiscal powers, local government stood for nothing more than a ceremonial transfer of authority to the local level. In recent years, most of the political parties have accepted devolution of power to local government, but their elections are not regularly held and they have not yet taken root in Pakistan's political culture. Disbursement of aid in the absence of these essential institutional mechanisms produced no results.

Success and failure of development assistance

Aid compatibility with institutional practices

Donors do not have resources to create the essential institutional infrastructure. They do not have sufficient knowledge to identify good practices and pioneers of social change and cannot afford to have patience to let the gradual but unpredictable processes of change produce the desired result. Their key concerns are scale, speed and time-bound results. As this does not happen by pumping in resources, they need to find out reasons. Since donors see nothing wrong in their approach, they find the explanation of their failure in 'lack of awareness' of the beneficiaries, 'lack of the capacity' of the government and lack of 'result-based management'. They conveniently ignore, for example, that millions of dollars spent on creating 'awareness' for education has led to a decline in the rate of enrolment; a large number of 'efficiently' delivered water supply schemes are abandoned soon after their construction; and Pakistan is home to thousands of ghost doctors and ghost schools. A second favourite choice of donors and NGOs is to demand signing and complying with global conventions and frameworks, without realizing the need for understanding current practices and investing in the processes to change these practices. These changes are possible through processes led by well-endowed local institutions. It is therefore important to look at some of the practices neutralizing the impact of foreign economic assistance. Let us begin with the government.

Government

Ineffectiveness of government and the donor-dependent CSOs arises in dealing with the universe of unwritten, unrecorded and undocumented social reality of low-income communities and marginalized groups; their living conditions; their human, social and financial assets; and the condition of existing social, administrative and physical infrastructure in their settlements. This unrecorded world can be transformed in the process of observing, recoding and transforming the existing practices with the participation of the people based on redefined roles and responsibilities. This world can be changed not through new templates but through new processes. Templates reproduce existing relationships. The process of accessing, creating and possessing documents empowers communities to deal with the inflexibility of templates and their incompatibility with the ground reality.

It is important to note here that successive governments in present-day Pakistan ignore the need to connect with local reality despite many policy decisions to do so. In 1952, for example, Government of Pakistan received assistance from the United Nations for formulating a national social policy. This was reflected in the first Five-Year Plan (1955–1960), and a social welfare section was created as a permanent section of the Planning Board, which was later changed to the National Planning Commission. In 1956 the National Social Welfare Council and subsequently provincial Social Welfare Councils were formed. The second and third

Five-Year Plans promoted community development by reaching out to people and involving them in development tasks (UNDP: 1991: 2–7). These reforms did not accompany any meaningful support to strengthen civil society.

Underdeveloped social infrastructure has led to ineffective management of public enterprises. The supplement to the Medium Term Development Framework on rural poverty reduction clearly states,

> The voice of Pakistan's poor captured in 49 districts from 121 community organizations during consultation for the Pakistan Poverty Reduction Strategy Papers tells us that this situation does not arise due to a lack of resources. Resources that are available for the public are not being used according to people's priorities in an efficient, effective and transparent manner.
>
> (GOP: 2005)

A review of the multi-donor supported Social Action Programme (SAP) by SPDC noted that "Expenditures on the social sector over the four-year period (1993/4 to 1996/7 aggregated to over Rs. 163 billion. Starting from a level of Rs. 27.7 billion in 1993/4 they have grown rapidly at the rate of 25 percent per annum" (Sattar: 2011). During the same period, however, the gross enrolment rate in primary education remained the same, full immunization of children doubled from 25 percent to 54 percent, and contraceptive prevalence rate doubled from 7 percent to 14 percent. A survey of selected villages during the same period undertaken by strengthening participatory organizations (SPO) reported dismal performance in provision of basic services. According to the survey, basic health units (BHUs) and rural health units (RHUs) existed in 33 percent of the communities surveyed; schools and BHUs were generally found in poor condition; only 48 percent of the male doctors were present and 18 percent of lady health workers (LHWs) were living in the communities surveyed. Inoperative water supply schemes were 40–50 percent of the total schemes. The review related this low performance to weak management, lack of community participation and abuse of discretionary power by government authorities (Sattar: 2011: vi).

At the local level, most government authorities do not have maps, data and documentation on basic services available in areas under their jurisdiction. Their budgetary system is a single-entry system that can only track the cash flow and not the resources available based on comparative statement of assets and liabilities. Our conventional planners do not understand the socio-economic reality of the poor. Their technical specifications for service delivery and infrastructure development in low-income areas are not in harmony with the economic condition of the poor. Standard operating procedures (SOPs) of government departments are cumbersome and obsolete. Implementation of government schemes at the local level by contractors considerably increases the cost of delivery. There is little understanding among the conventional professionals on how ordinary citizens can help improve our social sector's performance. These practices in tandem with the mindset which values "confidentiality" in use of public funds, discretion in use of rules and indignation over people's participation in decision-making, financing and managing local development supports and sustains inertia in social sector.

The Daudzai project undertaken by Pakistan Academy for Rural Development (PARD) in 1972 and Orangi Pilot Project (OPP) partnership with government agencies for community-based sanitation in Sukkur, Hyderabad and Karachi from 1982 onwards have offered valuable insights about the hurdles in the way of effective performance of government. Review and evaluation of these projects has shown that it is not lack of capacity but the existence of unguided, unused and uncoordinated capacity that explains poor delivery in low-income urban and rural settlements. (Khan: 1980; Hasan: 1997: 86).

The supplement to MTDF notes that

> The government has limited outreach to the rural areas and its institutional mechanisms and resources are limited. In order to focus development in rural areas it is essential to decentralize the planning and implementation of services to the local level and to encourage local communities and non-state sector.

It further added, "Organized communities provide the mechanism to overcome many of the governance problems faced today – from local level planning, to management and implementation, to monitoring" (GOP: 2005). Hasan has also noted that formal sector professionals do not have the capacity to plan from the point of view of poor and that leads to failure of many government and donor supported projects for community development (Hasan: 1997: 12–17).

Civil and uncivil society

A close review of Pakistan's economy shows that weak human development goes hand in hand with underutilized financial resources in the public sector (Baqir: 2007), existence of a large philanthropic sector (AKDN: 2000) and a narrow taxation base. Due to inaccessibility of services through the formal sector, low-income communities approach the informal sector for survival (Rahman: 2013; Hasan et al.: 2015). We need to note two dimensions of the problem of social capital formation in this regard. First, civil society following the formal sector approach has not helped communities appropriate the legal and fiscal space conceded by the state for full realization of human development potential. Communities have accessed these resources through "uncivil society", anchored in the informal sector. Second, the existence of the informal sector in large swathes of Asian, African and Latin American economies with sizeable assets is not integrated with the mainstream economy (De Soto: 2000; Anzorena et al.: 1998; Mustafa: 2005; Sattar: 2011; Hasan et al.: 2015). This is not only the state's failure but also civil society's failure. This requires a deeper probe into the cases of success and failure of civil society organizations dealing with the issues of community access to public resources (Leitner: 1882; Najam: 2000; Baqir: 2007). OPP's component sharing model provides one modality for integrating formal with informal sectors. There are many other instances of integrating formal and informal sectors from Pakistan. Without learning from such experiments, civil society cannot re-appropriate the space from uncivil society to further develop the receiving mechanism.

An overview of CSOs in terms of their outreach and source of funding shows that NGOs raising funds through user fees and indigenous philanthropy are playing a much greater role than foreign-funded NGOs because their models are based on better understanding of ground reality. They provide good examples of private-public partnership and the informal sector's integration with the formal sector. Social enterprises have done outstanding work in providing access to public resources to low-income communities through private-public partnership. At present, most NGOs in Pakistan fall in the category of *alleviating* the situation of the poor. According to Sattar, a large number of NGOs in Pakistan are engaged in service delivery, and among them madrassas constitute the highest percentage. The number of madrassas increased from 700 in the 1980s to 27,000 in 2010 (Sattar: 2011: 9). Formal sector, foreign-funded NGOs play a small role in community development despite their high profile. Most of the NGOs generate resources through fees and user charges (34 percent), followed by indigenous philanthropy (37 percent), foreign funding (6 percent) and public-sector support (6 percent). The National Commission on the Status of Women (NCSW) acknowledged the critical role of CSOs as intermediaries between women, marginalized communities and the state (NCSW: 2012: 77). The government is supportive of locally funded NGOs filling service delivery gaps and apprehensive of foreign funded NGOs working independently and engaged in advocacy on sensitive issues (ADB: 2015).

According to some experts, NGOs do not engage with government because they are afraid of backlash and repercussions (Sattar: 2011). Given the lack of NGOs' financial dependence on government, it is difficult to understand their fear of backlash. I think it is more due to a lack of capacity – for social preparation and engagement with the government – than a fear of backlash. My contention is that "means do exist for the population to free themselves, but the option has remained undeveloped" (Sharp: 2011: 13) – that is, the weakness of civil society and donor-funded NGOs engaged in poverty alleviation and right-based development. Inexperience, more than fear, explains NGOs' limited success in appropriating the fiscal resources from the government. OPP's experience has shown the use of mapping and documentation as a tool for advocacy on improvement of urban services. Mapping documents what already exists on the ground (in terms of sanitation infrastructure) and influences the government to align its investments with what already exists.

Communities

Formal sector experts see the failure of the state as failure of communities. Difference in the living conditions and strategies of survival of these communities is not seen as an expression of diversity but as lack of awareness, assets, finances and power of the people. Based on this vision they consider it 'their burden' to provide them handouts and subsidies and guide them to seek their rights. They do not consider it important to gain understanding of the people living in poverty and suffering from discrimination. They look at their lack of knowledge of people as illiteracy or ignorance of people (Chambers: 1997).

They do not even know that low-income communities have successfully established their own water, sanitation, shelter, health, education and job creation systems. They meet their daily needs with their own resources and manage to survive under very adverse conditions. Limited capacity of the formal sector to help the poor can be gauged by looking at the size of the unspent social sector budget of the government, the number of ghost facilities in education and the health sector and the magnitude of unutilized or misappropriated funds in the public sector, and the size of remittances by overseas Pakistani workers which is many times the size of official development assistance (ODA) coming to Pakistan.

They have no explanation for the fact that while people are willing to pay tribute, extortion money and security charges to local mafias, they do not pay taxes to the governments. They cannot explain why the poor prefer the 'illegal' informal practices to rational and formal practices. Why has the development discourse disappeared between the illegal and the rational? What explains the difference between the success or failure on any intervention for uplifting people is not the origin of these efforts in the public sector, market or civil society, but whether or not it enhances the range of choices of the people; whether this effort enables people to make the best use of their existing resources or limits their options to rise above their living standards. Why could the opportunities created by devolution of power to local governments in Pakistan not be realized? There is only one fact that explains it: lack of essential infrastructures and consequent hijacking of devolution by the local elite (Ahmad: 2008).

Conclusion

Pakistan's economy has received generous external economic assistance and at the same time underutilized financial resources in the public sector (Baqir: 2007). Pakistan has a large philanthropic sector – double the size of public sector development funds (AKDN: 2000) – and a narrow taxation base. Pakistan has thousands of abandoned health and education facilities and a large informal economy endowed with a vibrant asset base (Rahman: 2013; Hasan et al.: 2015; De Soto: 2000; Anzorena et al.: 1998; Mustafa: 2005; Sattar: 2011; Hasan et al.: 2015). Pakistan's key problem is not availability but efficient and effective use of financial resources. Governance in Pakistan is characterized by inflexibility, rigidity and red tape, and it has elicited two responses: uncivil society and social entrepreneurship. Social entrepreneurs and uncivil society both have succeeded by accessing public resources for community development; one by demonstrating the most effective way of using private means for achieving public goals, and the other by engaging government functionaries by offering part of the public means for private goals. If NGOs and government do not follow the flexible, realist and entrepreneurial approach, they will leave no option for the poor other than rallying behind the uncivil society. There are no shortcuts. Shortcuts lead to appropriation of public means for achieving private goals.

For this purpose, the key imbalance between the state and communities needs to be addressed. These are imbalances in the realm of political, fiscal or social participation. As a general practice, very few policy makers, planners and development practitioners try to make comparative analysis of government's failure and informal sector's success in solving problems related with poverty alleviation. The discrepancy between the performances of both sectors draws attention to the knowledge gaps of formal sector professionals in dealing with poverty and improvement of governance. Informal sector solutions are rejected because of the "poor quality" of those solutions. The way forward is upgrading the quality not rejecting the solution. In-depth analysis of these solutions, the problem-solving approach of their proponents, and the scale on which they were implemented shows that a poor person's solution to his/her problems needs to be minutely analyzed to draw lessons for formal sector planning for the poor. The knowledge deficit cannot be overcome by demanding more financial resources and a greater share in power by civil society but by gaining better understanding and knowledge of local priorities, tools, and strategies that can help deprived communities effectively meet their priorities. In the process, knowledge-based strategies lay the foundation for improved governance.

This book is an effort to understand where things go wrong when all the ingredients are available. I have tried to explore the solution to the problem of missed opportunities in the light of five stories from the field in Pakistan.

Notes

1   See Aga Khan III (1954). *The Memoirs of Aga Khan: World Enough and Time*, Cassel & Company, London.
2   The Village Agricultural Industrial Development Programme was launched in 1953–54 in East and West Pakistan with technical assistance from the US government.
3   *Markaz* means centre. It refers to a central office established at the Thana level (lowest tier of police administration) to bring different line departments under one "roof" to enhance their accessibility to the target beneficiaries.
4   AKRSP succeeded in doubling the incomes of its 100,000 beneficiary households in 10 years. It received the attention of the prime minister of Pakistan and chief ministers of various provinces as well as many bilateral and multi-lateral donors. Subsequently the Sarhad Rural Support Programme (SRSP), National Rural Support Programme (NRSP), Balochistan Rural Support Programme (BRSP), Sindh Rural Support Organization (SRSO), Thar Deep Development Programme (TRDP), Ghazi Barotha Taraqqiati Idara (GBTI) and Punjab Rural Support Programme were established based on AKRSP's model. UNDP provided funds to replicate this programme all over South Asia as the South Asia Poverty Alleviation Programme (SAPAP).

References

Aga Khan Foundation Report, 2000, *Philanthropy in Pakistan*, AKDN Pakistan.
Ahmad, M., 2008, "State of the Union: Social Audit of Governance in Union Council Bhangali 2007", *Lahore Journal of Policy Studies*, Vol. 2, No. 1, pp. 43–67.

Alavi, H., 1972, "The State in Post-Capitalist Societies: Pakistan and Bangladesh", *New Left Review*, Vol. 74, No. 1, pp. 59–81.

Ali, Chaudhri M., 1967, *Emergence of Pakistan*, Columbia University Press, New York.

Anzorena, J., J. Bolnick, S. Boonyabancha, Y. Cabannes, A. Hardoy, A. Hasan, C. Levy, D. Mitlin, D. Murphy, S. Patel and M. Saborido, 1998, "Reducing Urban Poverty: Some Lessons from Experience", *Environment and Urbanization*, Vol. 10, No. 1, pp. 167–186.

Asian Development Bank (ADB), *Civil Society Brief on Pakistan 2015*, viewed on December 11, 2015, from www.adb.org/publications/overview-civil-society-organizations-pakistan.

Baqir, F., 1984, *Process of Capital Accumulation in Pakistan: Achievement of Economic Goals by Political Means in a Market Setting*, Master's Thesis, Department of Economics and Business Administration, University of Idaho, Moscow.

Baqir, F., 2007, *UN Reforms and Civil Society Engagement*, UNRCO, Islamabad.

Beck, T., 1994, *The Experience of Poverty: Fighting for Respect and Resources in Village India*, Intermediate Technology, London.

Chambers, R., 1997, *Whose Reality Counts? Putting the First Last*, Intermediate Technology, London.

De Soto, H., 2000, *The Mystery of Capital: Why Capitalism Triumphs in the West and Fails Everywhere Else*, Basic Civitas Books, New York.

Easterly, W., 2001, *The Political Economy of Growth without Development: A Case Study of Pakistan*, Paper for the Analytical Narratives of Growth Project, Kennedy School of Government, Harvard University, Cambridge, MA.

Faure, G.O., 2011, "Practice Note: Informal Mediation in China", *Conflict Resolution Quarterly*, Vol. 29, No. 1, pp. 85–99.

Golub, S., 2003, *Non-state Justice Systems in Bangladesh and the Philippines*, DFID, Berkeley, CA. Online viewed from www.gsdrc.org/docs/open/ds34.pdf.

Government of Pakistan, 2005, *Medium Term Development Framework (2005–2010) Supplement: Rural Poverty Reduction through Social Mobilization (2005–2010)*, Planning Commission, Islamabad.

Haq, M.U., 1966, *The Strategy of Economic Planning: A Case Study of Pakistan*, Oxford University Press, Karachi.

Hasan, A., 1997, *Working with the Government: The Story of OPP's Collaboration with State Agencies for Replicating Its Low Cost Sanitation Programme*, City Press, Karachi.

Hasan, A., N. Ahmed, M. Raza, A. Sadiq, S.U. Ahmed and M.B. Sarwar, 2015, *Karachi: The Land Issue*, Oxford University Press, Karachi.

Islam, N., 1981, *Foreign Trade and Economic Controls in Development: The Case of United Pakistan*, Yale University Press, New Haven, CT.

Jamal, H., 2017, "Impact of Indian Dams in Kashmir on Pakistani Rivers", *About Civil Engineering*, viewed from www.aboutcivil.org/impact-of-Indian-dams-in-Kashmir-on-Pakistani-rivers.html.

Khan, A.H., 1996, *Orangi Pilot Project, Reminiscences and Reflections*, Oxford University Press, Karachi.

Khan, M.A., 1967. *Friends Not Masters: A Political Autobiography*, Oxford University Press, Oxford.

Khan, S.S., 1980, *Rural Development in Pakistan*, Vikas Publicizing House Ltd, Ghaziabad.

Leitner, G.W., 1882, *History of Indigenous Education in the Punjab Since Annexation and in 1882*, Calcutta.

Lewis, Stephen R., Jr., 1969, *Economic Policy and Industrial Growth in Pakistan*, Allen & Unwin, London.

Lewis, Stephen R., Jr., 1970, *Pakistan: Industrialization and Trade Policies*, Oxford University Press, Oxford.

Martinez-Vazquez, J. and M.R. Cyan, 2015, "Pakistan's Enduring Agenda for Tax Reforms", in *The Role of Taxation in Pakistan's Revival* (pp. 1–74), Oxford University Press, Oxford.

Mustafa, D., 2005, "(Anti) Social Capital in the Production of an (un) Civil Society in Pakistan", *Geographical Review*, Vol. 95, No. 3, pp. 328–347.

Myrdal, G., 1968, *Asian Drama, an Inquiry into the Poverty of Nations* (Vol. 1), Twentieth Century Fund, New York.

Najam, A., 2000, "The Four C's of Government Third Sector-Government Relations", *Nonprofit Management and Leadership*, Vol. 10, No. 4, pp. 375–396.

National Commission on Status of Women (NCSW) Annual Report, 2012, Opal Studio, Islamabad.

Naviwala, N., 2016, *Pakistan Education Crisis: The Real Story*, Wilson Center, Washington, DC.

Papanek, Gustav F., 1967, *Pakistan's Development: Social Goals and Private Incentives*, Harvard University Press, Cambridge, MA.

Radjou, Navi, Jaideep Prabhu and Simone Ahuja, 2012, *Jugaad Innovation: A Frugal and Flexible Approach to Innovation for the 21st Century*, Jossey-Bass, San Francisco.

Rahman, P., 2013, *Talk at Asian Coalition of Housing Rights' (AHCR) Gathering of in Bangkok "we are all Ninja Turtles of Mapping"*. ACHR Website, ACHR, 3.

Rapley, J., 2007, *Understanding Development: Theory and Practice in the Third World*, Lynne Rienner, Boulder, CO.

Sachs, Jeffrey D., 2005, *The End of Poverty: Economic Possibilities for Our Time*, Penguin Books, New York.

Sattar, N., 2011, *Has Civil Society Failed in Pakistan? Social Policy and Development Centre (SPDC)*. Working Paper No. 6.

Sharp, G., 2011, *From Dictatorship to Democracy, Serpent's Tail*, London, Shirkatgah, viewed on December 2, 2015, from http://shirkatgah.org/documentation-centre.

Sirolli, E., 2012, *Want to Help Someone: Shut Up and Listen*, Ted Talk. Online viewed from www.ted.com/talks/ernesto_sirolli_want_to_help_someone_shut_up_and_listen.

Tung, Mao Tse, 1926, *Analysis of Classes in Chinese Society-Selected Works of Mao Tse Tung*. Revised in HTML by Maoist Documentary Project at Marxist.org.

UNDP, 1991, *NGOs Working for Others: A Contribution to Human Development*, Vol. 1, United Nations Development Programme, Islamabad.

World Bank, Social Development Unit, 2008, *Justice for the Poor Program, Forging the Middle Ground: Engaging Non-State Justice in Indonesia*, pp. 6–9, 15–57. Online viewed from http://documents.worldbank.org/curated/en/2008/05/11242053/forging-middle-groundengaging-%20non-state-justice-indonesia.

2 Access to water and the science of financial patronage

"The government says there are water shortages," said Abdul Samad, resident of the poor Metroville area. "But we see tankers in our neighbourhood every day – where's that water coming from?"

(Dawn: 2015: 2, 6)

Karachi

It is interesting to note that while the problem of water – as described by Abdul Samad – appears clearly to be a problem of mismanagement to ordinary consumers in Karachi, it has skipped the attention of donors, bureaucrats and consultants for decades. It is therefore important to take a look at the resource endowment and water management practices to have a clear understanding of the problem. Karachi is a city of 20 million people and is home to 10 percent of Pakistan's population. It is the main port city of Pakistan as well as its financial, commercial and industrial hub. The city is spread over an area of 3,527 square kilometers. Half of the city's population consists of service providers, industrial workers, daily wage earners, home-based entrepreneurs, street vendors, domestic helpers and low-income government employees. These low-income workers live in 500 informal settlements. People in search of jobs, education and small businesses flock to Karachi in big numbers. That is why Karachi is known as the 'mother of the poor'. A city with a population of a few hundred thousand at the time of independence in 1947, Karachi has grown to the size of a metropolis in the past seven decades. Its diversity finds expression in pluralism as well as in hostility. It is called 'small Pakistan' because people from all ethnic and religious groups and geographic regions have come together as members of one big community in the city. At the same time the city is the turf of ethnic, sectarian and political violence and is dominated by criminal gangs.

The chief justice of Pakistan, in his judgement in a *suo motu* case on prevalence of violence in Karachi, gave the verdict that all the leading political parties and religious groups had militant outfits attached to them, and they are engaged in murder, extortion and ransom-taking (Zulfiqar: 2011). The use of violence for control over public spaces, assets and services takes place in the informal sector, and resource grabbers charge a fee for use of these resources from low-income communities not well protected by the law and order institutions. The prevalence of the public resourcegrabbing is closely connected with lack of transparency, exclusionist governance and interrupted rule of elected local government. Integrating the informal sector with the formal sector can take place through service delivery programmes based on community-government partnership known as a component sharing approach.[1] This can only be done by empowering local government and support organizations who can steer this process in collaboration with local partners (USIP: 2016: 3).[2] The key dilemma of the low-income communities in Karachi is that their contribution to the city through the development of infrastructures and service provisions in the informal economy is not taken into account. It is only through government's recognition of these assets and partnership with organized communities that the residents can get rid of the mafias and gangs claiming their 'share' in the 'management' of the city. Karachi does not suffer from lack of resources; it suffers from lack of good governance and sound management. Only the residents of Karachi can fix it, and no one else can do it *for* them.

Karachi's economy

Formal economy

Karachi is endowed with ample resources. Karachi houses 10 percent of the total population of Pakistan and 23 percent of its urban population. Its current rate of growth is estimated at around 5 percent, of which 3 percent is due to natural increase and 2 percent due to migration from other parts of the country. Its contribution to the national economy is much greater than its proportion of population. Karachi provides 25 percent of federal revenue and 15 percent of Pakistan's gross domestic product (GDP). In addition, 50 percent of the country's bank deposits and 72 percent of all issued capital is contributed by the city (Ahmed: 2010: 120). It is difficult to understand how a city so well resourced cannot deliver services due to lack of finances and needs to resort to international borrowing to solve its basic problems. If the resources contributed by the informal economy are taken into consideration, the justification for borrowing is further weakened.

Informal economy

The case for lending by international financial institutions (IFIs) for improvement of service delivery is weaved around the myth of poverty. Karachi's poor is not textbook poor, and much more dynamic, resourceful and brilliant than its pale image painted in the project documents of development assistance agencies. They have purchased each and every inch of land they inhabit from the land grabbers.

Asset value of land under informal settlements in Karachi has been estimated to be Rs. 500 billion (Sayeed et al.: 2016). It is very much in line with the global estimates of the financial value of land owned by residents of informal settlements. Hernando De Soto has asserted that "Property value of the land held by the squatters is greater than the value of all major stock exchanges in the world" (De Soto: 2000: 32–33). Half of Karachi's population depends for its livelihood on the informal sector and receives all the civic services and amenities through the rent paid out to various extortionist mafias. While the poor pay for all the services they receive, the government does not receive what the poor are willing to pay. This creates the structural foundation of the financial patronage system. By refusing to accept service charges from the poor, the governance system misses the opportunity to establish reciprocity between the service providers and service receivers and leaves low-income communities at the mercy of discretionary power of the administrators. Service delivery under such circumstances takes place through local 'power brokers' who serve as clients of their political and administrative patrons.

At this point it is important to mention that the poor of Karachi daily pay Rs. 830 million (US$7.9 million)[3] to various mafias in the city in terms of receiving basic services, protection and security, loss of valuables through theft and extortion and ransom to criminal gangs. This is equal to one ADB loan daily. They pay Rs. 10 million to local gangs as security money; Rs. 50 million as ransom to kidnappers; Rs. 2.4 million as 'fees' to 500 illegally established parking lot operators; and Rs. 8.25 million is paid in extortion by 55,000 street vendors. The water mafia collects Rs. 100 million for selling 272 million gallons of water illegally. Rs. 150 million is collected in bribes by police from 15,000 drug-selling and gambling dens. Karachi's land mafia illegally grabs over 30,000 acres of government land, denting the national exchequer by Rs. 7 billion annually. The city's transport mafia extorts Rs. 10.48 million from buses, rickshaws and taxis. Illegal payments are called 'Dhakka Wasooli'. Rent collected from trucks, container trucks and oil tankers bring the mafia an additional Rs. 7.5 million. Electricity worth Rs. 10.5 million is stolen daily in the city through 4,000 to 5,000 *kundas* (illegal connections). Criminals associated with the health industry deprive poor patients of Rs. 3 million every day. On average, 40–50 motorcycles and 20–25 cars worth Rs. 20.5 million are lifted daily. Street criminals loot around Rs. 5.2 million daily by snatching mobile phones, cash, jewellery and other valuables. Short-term kidnapping, in which the abductees are held at gunpoint and driven around the city for several hours, generates over Rs. 3 million. The police pockets Rs. 210 million daily in bribes (AFP: 2013: 1).

Karachi's poor

As very aptly described by Akhter Hameed Khan, the poor should not be confused with the destitute. They pay for all the services they receive. They don't want handouts. It hurts their dignity and pride. The problem of the poor is partnership, not patronage. They have built their own assets and infrastructures and they survive because they are connected with these service delivery systems. They are not treated fairly. As very eloquently stated by Bryan Stevenson, the opposite of poverty is not prosperity; it is justice (Stevenson: 2012). The poor pay for all the services they receive but the formal economic, political and justice systems do not have the capacity to deliver to them or integrate their informal delivery systems in the formal economy.

Water problem in Karachi

Karachi's poor have been deprived of access to clean drinking water for a very long time. The issue drew the attention of the Karachi Water and Sewerage Board (KWSB), federal, provincial and local government authorities, development assistance agencies, bilateral and multilateral donors, national and international NGOs, and international financial institutions (IFIs) in the 1990s.[4] It is very interesting to note that most of the players looked at the problem from the perspective of their own institutional interests, being completely oblivious to the situation on the ground and the best way of fulfilling the unmet need of the low-income communities.

To use the deprivation of a neglected section of society as an opportunity to hand out grants and 'technical advice' and sell loans, many institutions spun the myths of the underdeveloped professional capacity of the KWSB, the shortage of funds for building physical infrastructure, and the 'poverty' of the residents of informal settlements. This type of development assistance, which is not based on the understanding of social reality but the 'delivery' needs of lending and grant making institutions and unjustified fiscal decisions of the recipient institutions, constitutes a system of financial patronage which perpetuates the status quo in the name of sustainable development or poverty alleviation. A close view of the situation on the ground shows that it is not the poor of Karachi who need the help of IFIs to have access to water, but it is the IFIs who need to learn from the poor to withdraw their expensive, irrelevant and ineffective solutions. Citizens' rights can neither be purchased with expensive foreign loans *on behalf of the poor* nor achieved by raising the sizzling slogans on the streets inspired by the rhetoric of 'international obligations', but by sitting at the negotiating table with the government with documented evidence of the ground reality. This is what worked in Karachi. That is what contributes to clear understanding of the options for finding *nonzero* and sustainable solutions.[5]

The development assistance industry includes grant-making institutions like UN agencies, bilateral donors, regional and global loan-making financial institutions and international NGOs. Their aid objectives may be different but their approach to provision of services is similar to the formal sector professionals of the donor and aid recipient countries. People being 'helped' through grants and loans live under the informal economy. Without understanding the difference in the social conditions of people living in the informal sector, the aid aimed at improving their situation leads to unintended negative consequences.

For making the distinction between the formal and informal sectors clear, I will present the formulation of problems from both the sources. Many development assistance agencies have offered technical assistance and infrastructure development programmes to solve water supply and other urban service delivery issues in Karachi since 2006. Prominent among these agencies are the World Bank, Asian Development Bank (ADB), Japanese Bank for International Cooperation (JBIC), Japan International Cooperation Agency (JICA) and Japan External Trade Organization (JETRO; Hasan: 2006: 2). The assistance offered ranged from preparation of plans for future urban development and provision of technical assistance by expat experts to investment in construction of physical infrastructure and improvement of urban basic services including water, sanitation, waste disposal and mass transit. A major weakness of all these projects was that they were not in line with the ground reality. Ground reality is captured in Figures 2.1 and 2.2. It resulted in cancellation of one loan and the lack of any progress in water supply to the poor through other projects. How so much aid resulted in so little accomplishment can be understood by taking into consideration the ground reality.

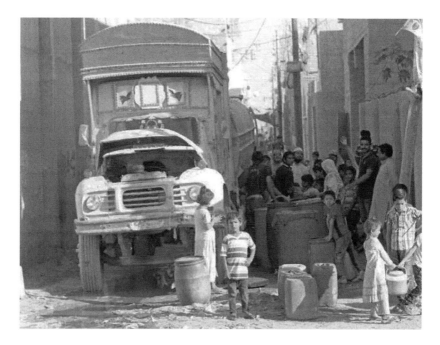

Figure 2.1 Water tanker

Source: Photo by Ayesha Saleem, *Express Tribune*, August 25, 2016

Figure 2.2 Water tankers

Source: *Express Tribune*, June 28, 2017

Water leakages and theft

Let us now take a snapshot of water services in the city. Does the city suffer from a water shortage, resource deficit or inability of poor customers to pay the water service charges? Or does the problem lie in managerial negligence and poor governance of the Karachi Water and Sewerage Board (KWSB)? We are very fortunate to have reliable evidence on all these aspects of drinking water supply in Karachi and we can review them one by one (Ahmed and Sohail: 2003; Askari: 2015; Hasan: 2005b). A newspaper report pointed out that about 45 percent of the total supply of water was being siphoned off due to illegal connections and leakages in the obsolete pipelines. Officials turned a blind eye to this loss. At one pumping station, a 15-inch pipe line had developed a wide leak and water was going into a sewerage line; area residents reported the situation to KWSB and no remedial measures were taken for about 15 years (Dawn: 2007). Officials in the KWSB themselves acknowledged the existence of 150 illegal hydrants drawing water from its main pipelines causing an acute water shortage in the city and a massive revenue loss of over Rs. 1.3 billion annually to the KWSB. A senior KWSB official said on condition of anonymity that "A large scale theft of over 30 million gallons of water occurs each day from the KWSB's various pipelines. This is mostly done in connivance with low ranking staff of the KWSB and the concerned area police" (Dawn: 2011).

In a workshop organized by the World Bank and UNDP-funded Water and Sanitation Programme (WSP) in 2005, the experts endorsed the views expressed above by pointing out that "Improving water and sanitation service delivery in South Asia's growing cities is not about fixing the pipe – it is about fixing the institutions that fix those pipes". The report also pointed out that "unaccounted water in the region is over 40 to 50 percent and only 20 to 30 percent of the operation and maintenance cost in the water sector is being recovered". Water losses in transmission are 35 percent in Karachi (WSP: 2005: 1). They also pointed out that "the network map shows the connections and makes it possible to monitor water supply and total billing" (WSP: 2005: 9). The workshop pointed out that reliable information, awareness of choices and strong political leadership could play a critical role in solving this problem with the engagement of citizens. The participants noted that poor are already paying a high cost for water'and proposed that "local revenue generation based on economic pricing of water – is the approach to follow" (WSP: 2005: 13). This was endorsed by a local NGO, the Hisaar Foundation, which found out that around 40 percent of the water in the city is wasted due to leakages in the supply lines (Dawn: 2011: 2). A newspaper report in 2016 mentioned that KWSB had detected 14 legal and 196 illegal hydrants in the city illegally selling 7 million gallons of water daily (*Daily Times*: 2016: 1).

Billing and metering

Drinking water supplied by KWSB is billed but is neither metered nor recovered in most of the cases. Lack of water metering results in wastage; it also leads to applying the same rates for supplying water to the poor and higher income brackets. Lack of recovery causes financial woes to KWSB. According to Ameena Batool, a housewife from eastern Karachi, "We haven't been getting water through the official supply for many years, although I regularly pay the KWSB water tax". In order to meet the daily water needs of the household, she pays about Rs. 500–600 a week for purchasing water from a tanker (Dawn: 2011). A large number of customers (38 percent) reported not receiving any water bill for water consumption (WSP: 2010: 31). This complaint was shared by the city *nazim* (mayor), Syed Mustafa Kamal, who said that KWSB's billing system in the past was defective and only 0.2 million out of 1.6 million bills used to be distributed (Dawn: 2007). But the situation did not improve even 10 years after the nazim took notice of the problem. Over 40 percent of the customers facing problems approached authorities through their elected councillors while 60 percent of the low-income customers never complained. Over 47 percent of the complainants who approached KWSB found it inaccessible (WSP: 2010: 132). This lack of response to customer's complaints is not due to lack of staff, because KWSB's administrator complained of overstaffing as one of the source of financial liabilities of KWSB. While lack of billing accounts for major financial deficit of KWSB, lack of recovery is an equally important cause of financial problems.

Levying and recovery of water charges

It is very interesting to look at the list of defaulters. No foreign 'technical expertise' is required to recover pending dues from these defaulters. It has been pointed out that

> A huge amount of over Rs. 5 billion – which is almost equal to that of the KWSB's annual budget – is outstanding under the head of water and sewerage charges against various departments and organizations belonging to a number of federal ministries such as defence, science and technology, commerce, food, agriculture and livestock, industries and production, etc.
>
> (Sharif: 2008: 1)

Collection is two thirds of the KWSB budget (WSP: 2005: 13). It is not deficient because people are not willing to pay. People from lower-income communities usually "end up paying 12 times the price for drinking water than what people from higher-income brackets pay" (Dawn: 2011). A WSP report noted that low-income communities paid as high as 40 times the charges paid by middle income groups. They pay Rs. 400–500 per month to purchase water from tankers (Rahman: 2004: 12). According to Iftikhar Ahmad Khan, the deputy managing director of KWSB, due to government crackdown 200 illegal water siphons were closed and tanker owners were compelled to buy water from KWSB at the cost of $1–2 per 3,700 litres but they resold this water at 10 times the cost in various neighbourhoods (Dawn: 2015).

Current system of access to water

WSP reported that due to the dysfunctional water supply system of KWSB, consumers approach vendors in the informal sector for purchasing water. These vendors include donkey cart and bicycle water carriers (6 percent); borehole or personal storage facilities (8 percent); mains connection or shared storage (10 percent); bottled water (13 percent); private tankers (20 percent); illegal mains connection with personal storage (37 percent); and mains connection with personal storage (38 percent). Only 45 percent of the respondents reported receiving water supply seven days a week. There was wide variation among towns on regular supply of water, ranging between 4 percent and 81 percent of the time (WSP: 2010: 23).[6] It compelled residents to purchase water from more expensive sources during periods of scarcity. Outside purchase was mostly done by the low-income customers.

> Tankers supply 20 percent of the water in the city, including to those who already have a KWSB supply to overcome the shortfall. Quite often they will pay anywhere from Rs. 6,000 to Rs. 10,000 for a month's supply, or about 30 times the rate the government charges. But people are glad to pay, as they have little choice.
>
> (Dawn: 2016)

Size of community investments and assets

Communities and CBOs are constantly engaged in investing in the development of water supply and sanitation infrastructure. According to Rahman, low-income communities have built up to 80 percent of the physical infrastructure with their own contributions inside their neighbourhoods. Since this work is of inferior quality, it needs to be upgraded, not replaced. These communities need to be provided technical assistance and managerial guidance, not expensive replacement of their assets. How such guidance can be provided and by whom is an important issue (Hasan: 2005: 2). Reform and improvement in public water delivery systems is ultimately determined by the level of effectiveness and accountability of the local government. In this context it is important to have a look at the evolution of the local government system in Karachi.

Government, governance and the poor

Case of KAWWS

A remarkable pioneering public interest litigation activity was carried out by a resident of the Karachi Administration Women Welfare Society (KAWWS), Safina Siddiqi. When the local administration did not comply with its mandate to provide sanitation services to residents of KAWWS, Safina in collaboration with a large group of housewives decided to approach the court. Since KAWWS was a formal settlement, its residents could claim the legal right to receive services against the taxes they paid to various tiers of government. A local judge allowed the residents not to pay taxes until the services were resumed (Hasan: 1998).[7] The services were reinstated. One of the consequences of integrating the informal economy in the formal governance system would be the entitlement of squatters to similar rights. In the absence of any leverage on local government, the poor approach mafias to meet their service needs.

Continuity, authority and capacity of local government

There are three issues faced by the local government in effectively dispensing its functions: continuity, fiscal authority and technical capacity. The Karachi municipal government was formed during the British rule in 1852. It was a non-representative body consisting of appointed members. It did a commendable job in developing the city, but it only enjoyed the trust of powers that be. It passed through various phases of reconstruction. During its 160-plus years of existence since its inception, the Karachi municipal government has been governed by elected representatives for 36 years: from 1964 to 1969 under Basic Democracy Order 1962; from 1979 to 2001 under Sindh Local Government Ordinance 1979 under non-party based elections; and from 2001 to 2010 under Sindh Local Government Ordinance (Ahmed: 2010: 124–125). There were breaks and discontinuities in the mode of governance of elected government even during this brief period. From time to time new institutions having overlapping functions with the municipal government were created, merged and restructured until the creation of City District Government Karachi (CDGK) in 2001 as an overarching civic body overseeing the functions of all development and planning agencies. The fiscal and administrative functions were not fully devolved to the CDGK and it depended on approvals and allocations from the federal and provincial government for its operation. Local bodies did not have credibility among political leadership because they were always used as a channel of political patronage during the ban on political parties under various military rulers. In the absence of elected heads of local government, the municipal government was headed by the bureaucrats. This arrangement continued until 1996 in Karachi.

In 1999 the CDGK headed by an elected representative was formed by bringing under one structure all the development agencies. In 2010 the elected regime was replaced by bureaucrats again. Karachi Development Authority (KDA), the predecessor of CDGK, raised revenue by land sales and development charges. Karachi Municipal Corporation (KMC), the second constituent of CDGK, raised revenue through collection of octroi taxes, property taxes, road tolls, conservancy taxes and grants from provincial and federal government. The planning wing of the city government and Environmental Control Department (MPECD) also went through many ups and downs during 1973–2000, and eventually their tasks were handed over to a private firm in 2003. Competing business, political and ethnic interests in governing the city are accommodated not through transparent, evidence-based negotiations but through informal arrangements (Ahmed: 2010: 123–130). During the early period planning decisions were taken by the provincial and federal governments, and this practice has continued even after formation of an elected local government. Limited fiscal autonomy together with extremely limited documentation of physical assets and infrastructures has led to an inherent gap between plans and local social, physical and administrative realities. Noman Ahmed has given a critical insight by saying that "planning is not a product. It is a continuous ongoing process" (Ahmed: 2010: 134). Continual interruption of this process has undermined the city government in properly performing its functions.

Local government authorities in Pakistan ranging from the Union Council to the CDGK manage their services without any mapping, documentation or demographic data.[8] Their revenue collection is based on outdated property assessments and a lack of recovery of service charges. There is also very limited coordination and knowledge sharing between the government departments and donors on preparation of local development projects[9] It is the case with CDGK as well. One wonders how the world-class consultants flown in for preparing project documents and loan agreements can assess the need for assistance, decide the design of the project and prepare realistic cost estimates. The IFIs' hurriedly developed project documents fail to offer the 'technical assistance' they promise to provide. This has repeatedly happened in Karachi and has also happened in the case of a loan agreement prepared by ADB for Karachi. CDGK does not have updated maps, property records or the latest census data (Ahmed: 2010; Hasan: 2005). This highlights the need for improvement of statutory provisions under Sindh Local Government Ordinance (SLGO) 2001 to make CDGK a body responsive to popular needs.

It is important to note here that continued existence of democratically elected governments is a necessary but not sufficient condition for an inclusive governance regime. In India, for example, the continuity of local government at all levels did not institutionalize pro-poor policies in provision of services to urban slum dwellers. This happens due to inherent anti-poor bias of the financing, planning and executing agencies. One wonders who benefits from such development assistance if it does not serve the people who are supposed to be its target beneficiaries. The answer is simple: IFIs are interested in meeting their loan disbursement targets; bureaucrats want air-conditioned offices, laptops, vehicles and foreign trips for their 'capacity development'; and political leadership wants foreign exchange to cover the foreign exchange deficit. Due to a lack of data and limited management capacity of the city government, it suits local government authorities to ask for 'new schemes' completely ignoring the infrastructure on the ground. This has become a popular practice in 'development business' and has been the practice followed by CDGK, the Government of Sindh and KWSB's request for a loan to 'fix the pipes', for the provision of water was no exception. The poor appear nowhere in this exercise (Hasan: 2012; Hasan: 2005). This practice has been checked only in those cases where local civil society organizations equipped with local knowledge, data and mapping were able to question the proposed IFI-funded project (Hasan: 2005a). In case there is no local resistance to such plans, the pattern keeps repeating itself, resulting in added loans and no improvement of the situation on the ground.

Deficiencies of these anti-poor projects cannot be overcome by provision of large-sized loans without aligning the approaches, systems, procedures, time frames and scale of funding of international agencies with the social reality of the poor. These projects do not relate to social reality because they were meant to lend money to governments, not to help the poor directly. Since IFI requirements do not permit investment in long-term and unpredictable processes, the only way they have been found useful by pro-poor development practitioners is in their use as a bridge fund by locally financed development projects between the periods of cash flow constraints. Meaningful investment for community uplift requires changing the framework and basic practices of IFIs, not provision of more funds on failed patterns (Hasan et al.: 2005a: 17).

Local governments can effectively respond to the citizens' concerns by progressively increasing the space for citizen participation not only in consultation for prioritization of issues, but also in budgeting and monitoring process and engaging them in co-governance of development programmes. It entails acceptance of the slow, messy and contradictory nature of the process that eventually outpaces the conventional professionally controlled processes during the course of implementation. This participatory co-governance also entails that citizen organizations not only assert rights but take responsibility for jointly finding innovative, information-based solutions (Hasan et al.: 2005a: 15). This moves CSOs from the picket line to the negotiating table.

As Hordijk has rightly pointed out, it is a "process of incremental learning for all parties involved" (Hordijk: 2005: 219). It cannot produce any results in a time-bound, result-based, project-oriented framework. In this process citizenship is not confined to the conventional sense of the right to vote or demand but assumes a larger role of "agency to act and practice". It needs a different kind of support mechanism as well. It does not fit well with the professional practices of IFIs because "Several IFIs depend on technical staff to deal with issues of social nature and therefore lose good opportunities for dialogue and building social cohesion" (Water Aid: 2009: 13). To depict the unique social nature of the water supply problem in Karachi and how the conventional 'technical assistance' shotgun missed this target, we need to take into consideration three key nodes of the water supply system and their number and practices: KWSB, CSOs and IFIs.

Nodes, numbers and practices

Karachi Water and Sewerage Board

Karachi Water and Sewerage Board (KWSB) was created by enactment of Sindh Local Government (Amendment) Ordinance of 1983. It was given the task of providing water and sanitation services in Karachi. The board was made autonomous in 1996 (WSP: 2010: 14–15). KWSB had no influence over determination of the tariff rate and there was no independent regulation on tariff in place. It had no relationship with customers and civil society organizations. A review of resolutions made by elected union councils revealed a low priority for water services. KWSB Managing Director Brigadier Iftikhar Haider in his 2007 presentation highlighted the need for institutional reform in his organization (Figure 2.3). He specially emphasized the need for rationalization and collection of tariff and human-resourced audit; costing charges, monthly billing, regular collection of bills, billing according to income level and nature of user (domestic, business and public service institution); consulting CSOs in deciding tariff rates and levying charges based on metering to overcome the imbalance between Rs. 34 billion monthly expenses and Rs. 2 billion monthly collection (Haider: 2007: 6). The solution was simple: metering, billing and collection of water charges.

Civil society organizations

OPP's architect Perween Rahman had developed a repository of local knowledge by mapping land use and infrastructure in 334 informal settlements (known as *katchi abadis* in Pakistan) by 2004. No government or donor agency had access to such a detailed source on the situation of urban poor in Karachi. Her findings on the nature of the problem of urban basic service delivery and its solution were based on real numbers. She pointed out that water shortages were caused due to leakages, unequal distribution and an ad hoc style of management. KWSB had no maps of the service area and it was not transparent (Rahman: 2004: 13). Findings of KWSB and OPP highlighted the need for stopping water theft and rationalizing

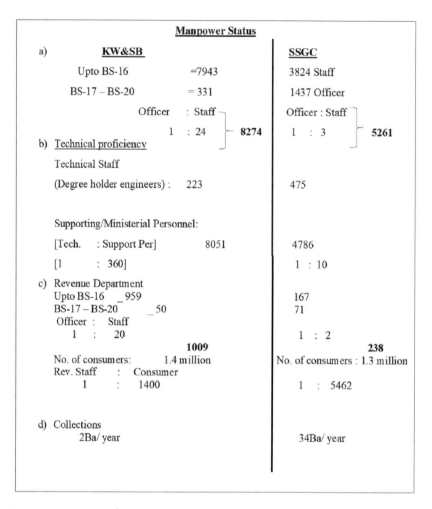

Figure 2.3 The imbalance

Source: Image of Slide 4 of presentation made by KWSB Managing Director Brigadier Iftikhar Haider on March 16, 2007. The comparison of KWSB and Sui Southern Gas Corporation's (SSGC) revenue collection through metering of services clearly shows that KWSB does not suffer from lack of resources but lack of sound management.

tariff to achieve Indicator 10 of MDG 7 on ensuring environmental stability to halve the proportion of people without safe water and basic sanitation between 1990 and 2015. No loan or 'technical assistance' for situation analysis, capacity building and 'demanding rights' was needed. All the ingredients of the solutions existed side by side with the problem. The only task needed was engaging and activating the government. This could be accomplished by helping the residents of informal settlements in negotiating with the government with the help of maps, numbers and calculations (Hasan et al.: 2005: 7). Rahman precisely undertook this task.

Rahman decided to map out the entire water supply route in the informal settlements. Her work showed that water supply to 500 squatter settlements in Karachi could be arranged without costing a single extra penny to the government's exchequer. She convincingly argued that the 665 million gallons daily (MGD) of water allocated to the city for drinking purposes was sufficient if bulk siphoning of water could be controlled. It is ironic that a well-resourced institution like KWSB could not stop water leakages. OPP's research showed that 12,000 tankers were stealing water from bulk mains and selling it for Rs. 49.6 billion (US$820 million, at US$1 = Rs. 60.5) per annum. KWSB (2007–2008) was incurring a loss of Rs. 2.8 billion per annum. Prevention of water theft could help KWSB easily cover its financial deficit and save as well (Rahman: 2008: 10–22).

Technical assistance of IFIs

The work undertaken by Rahman convincingly showed that reducing the cost and ensuring sustainability of service delivery projects depends on selection of a credible and competent local partner. Such a partner must have the capacity to document and manage local body of knowledge. Importance of this capacity was clearly demonstrated in the case of the Korangi Waste Water Treatment Project. OPP, Urban Resource Centre (URC) and their partners were able to convince the governor of Sindh, Moinuddin Haider, to cancel the loan because their estimates based on sound mapping of the project area were much lower than ADB's project cost based on the guesstimate (Hasan: 2005a: 20). Failure of most of the well-funded projects is due mainly to the absence of effective governance institutions. This raises the question whether the project funding can be used in establishing a "sustainable planning and management agency" for the city. Experience in Karachi shows that the managerial practices undertaken by the IFI-funded project are at odds with the processes needed for institution building, participatory development and social accountability. Hasan notes that as a result of loans taken from IFIs, water and sanitation services have not improved but "the Karachi Water and Sewage Board (KWSB) owes the ADB more than Rs. 42 billion" (Hasan: 2006: 1). It is therefore important to explore where the IFIs' approach is at odds with ground reality and where they need to change.

Development agencies' perception of the problem

The shotgun diagnosis – UNDP

Technical assistance or physical infrastructure investment support approaches undertaken by IFIs and other development agencies have two versions: shotgun vision and shortcut execution. UNDP was not engaged as a big player in dealing with the problem of access to water in Karachi, but UNDP's Annual

Report 2017 provides a good example of the shotgun vision on access to water. Under this vision every possible cause under the sun is put together as the list of problems to be handled, so that if one intervention does not work the failure can be blamed on the other missing inputs. See for example the following statement:

> While water related issues have been discussed as part of the National Climate Change Policy and National Drinking Water Policy, [a holistic national water policy is required]. Water pricing to promote efficient use of water, building water storage infrastructure to store excess water, enforcing strict water quality management systems to curb water pollution, controlling population growth and adopting a sustainable pattern of urbanization, are all major issues that require immediate attention if Pakistanis are to have access to the water they need in the future. Bold actions are needed to address this water crisis, otherwise not only will Pakistan not meet the SDGs on water, but its future development will be hampered.
>
> (UNDP: 2017)

It is interesting to note here that a community-based organization (CBO) which successfully turned around the situation at two bottom tiers of local government – Union Council (UC) 62 in Lahore and Town Municipal Administration (TMA) Bhalwal – without any donor assistance or government funding, summed up its strategy in one line: "water management is water measurement".[10] This simple formulation can provide deep insight into the problem and furnish clear and practical guidelines to make the change without changing the whole world.

There are inconsistencies in this shotgun approach perhaps due to the hurry in which it has been developed. While UNDP report calls for a holistic national policy, it forgets to include people in its version of good governance, which is supposed to be the principle means for achieving SDGs and improved water supply for the people in a holistic way. UNDP's Annual Report 2015 gives a conflicting message on this issue. Its vision of good governance does not go beyond good government. The report states:

> Strong, trusted institutions lie at the heart of a stable and equitable society. UNDP works with institutions at federal, provincial and local levels to build trusted, efficient and effective institutions – working with elected officials, civil servants, police and members of the judiciary – in service of Pakistan's people.
>
> (UNDP: 2015: 7)

There is no mention of community or civil society organizations. This might be based on the tacit assumption that people are part of the problem, not of the solution. This tacit assumption underlies all the development assistance projects designed by the development assistance fraternity.

The shortcut execution – ADB

The shortcut execution component promises to solve all the issues raised under the shotgun vision through the magical delivery of a short-term project. An example is offered by the ADB's technical assistance.

Project Number: 38405 Loan Number: 2229-PAK aimed to "speed up the development process in Karachi" through sequenced interventions. Under this project "Taking a long term and holistic approach" in contradistinction to "one off project approach" was sanctioned by ADB in 2006.[11] Size of the project was $10 billion, out of which the foreign exchange component was $6,460 million and the local component was $3,540 million. The project allocated $6.144 million for capacity building and $3,540 for project preparation facility. In actuality, both components of the project constituted 'technical assistance' to the 'client', which in this case was the city government of Karachi. The closing date of the project was 2009 (ADB: 2010: iii). ADB's own project evaluation considered it a failed project because "the project suffered from significant underutilization that first caused a partial cancellation of loan proceeds, followed by an almost complete project cancellation". The main cause for the severe underutilization was the lack of consensus between the various stakeholders at the provincial and CDGK levels on the scope of the proposed MFF (ADB: 2010: 4). Clearly the problem underlying defective delivery of water was not a financial capital deficit but a social capital deficit. ADB's approach to problem-solving seems to have serious shortcomings in dealing with the prevalent governance practices and social reality of the poor. This is also true about other IFIs. Scrutiny of the ADB project provides insights about the incompatibility of formal sector approach with the informal practices. Before discussing the formal sector approach in detail, it is important to mention the simple, inexpensive and sustainable solution presented by the architect of the poor, Perween Rahman, based on understanding of the local situation. Figure 2.4 gives an interesting visual depiction of the difference between the expert and community approaches to solving local development problems.

Field-based approach

Rahman's research showed that the poor of Karachi were paying a much higher price for water compared to the middle-class residents and closing down of illegal tankers would reduce their monthly water bills and effectively handle the financial woes of KWSB. Her work clearly showed that if all illegal hydrants were closed, KWSB could cover all its current expenses and save as well (Rahman: 2008: 10–22). This would have meant no justification of the loan and no need for any 'technical assistance'. However, the ADB loan was signed. It was cancelled after a couple of years due to noncompliance of the 'client' and at the federal government's request.

It is important to understand the dynamics of project development and its connection with the ground reality. ADB's own project evaluation concludes:

> The purpose of the TA loan (for Karachi Mega City Development Project) was to address the megacity development needs through a long-term and

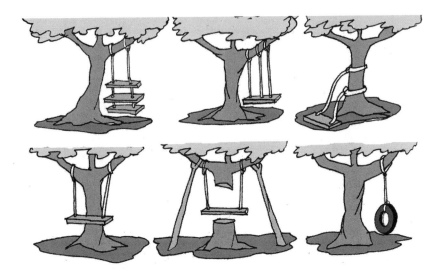

Figure 2.4 Perspectives
Source: Akhtar Shah

holistic approach by (i) building capacities within the city district government of Karachi (CDGK) and the various tehsil municipal authorities (TMAs) for effective city planning and management, (ii) adopting and implementing commercial principles in the provision of infrastructure and services, (iii) supporting the preparation of bankable projects for megacity development that might subsequently be funded by ADB through its loans to Karachi city, and (iv) promoting reforms for sustainable financing. The TA loan was intended to represent the first intervention of long-term assistance from ADB, and it was to establish the required foundation for effectively utilizing and sustaining future infrastructure investments in the megacity.

(ADB: 2010: 1)

How this diagnosis was carried out is anybody's guess. There was no documentation, mapping and research-based evidence available with ADB. The Orangi Pilot Project (OPP), a local NGO, was the only organization that had carried out mapping and documentation of local informal settlements over the past three decades. Even the City District Government Karachi (CDGK) and KWSB did not have the documentary evidence possessed by OPP. It would be of interest to note here that architect Perween Rahman, who worked on the diagnosis and solution of the problem based on her documentation of informal settlements, pointed out that the water problem could be solved by a knowledge-based and transparent management system (Rahman: 2004: 13). The diagnosis and solution of the problem were precise, simple, doable and within the reach of CDGK.

Water distribution map showing illegal hydrant

Rahman's survey also revealed that infrastructure development through international borrowing added to the burden of the government instead of reducing it. It added a crushing load on the poor because these loans and their interest are paid back through taxes. Sixty percent of the taxes in Pakistan are indirect taxes, and the poor bear the brunt of these taxes in paying back the ill-advised loans. As aptly summed up by Rahman, "85% of Pakistan's entire government budget goes into repaying the country's IMF and World Bank loans and to defense, so there is hardly anything left for the people, hardly anything left for development" (Rahman: 2013: 1). According to Rahman's calculations, the cost of building infrastructure by the government to provide water in informal settlements, taking into consideration water connections made by community on a self-help basis, was worth Rs. 435.18 million. If government carried out this work without building on community assets, its total expenses could go up to Rs. 630.94 million. If the same work had been done through an international borrowing, tendering and contracting process it would have amounted to Rs 2.57 billion – that is approximately six times the budget based on community participation and government contracting (Rahman: 2004: 12). Despite lavish funding, loans have not been able to make much difference. The ADB loan provides one example.

Explanation of the highly overdeveloped and lavishly financed ADB project's failure is given in the evaluation report in these words:

> The TA loan specifically aimed to establish a dedicated project preparation facility, and a specialized financing vehicle (SFV) to catalyze infrastructure investments. . . . The project's goal and impact were not achieved . . . the project suffered from significant underutilization. . . . The loan became effective on 14 September 2006 and was originally scheduled for closing on 14 April 2009. . . . On 5 November 2008, the Ministry of Economic Affairs and Statistics of the Government of Pakistan – the borrower – sent a letter to ADB, notifying that a large portion of the loan be cancelled without waiting until its closure on 30 April 2009.
>
> (ADB: 2010: 2)

It was further noted in the evaluation report that "The project's goal and impact were not achieved" and mainstreaming them was not considered feasible "within the confined [duration] of the TA loan" (ADB: 2010: 2). It clearly highlights the limitation of shotgun and shortcut approaches in solving the problem of service delivery to the 'poor' communities.

One of the biggest project risks which turned out to be true was "delayed implementation" of all the activities proposed under "technical assistance" (ADB: 2010: 3). In addition, the report noted that the project was too complex and not in line with ground realities (ADB: 2010: 8). It is important therefore to have a glance at the ground reality, which depicts that the people, government and service delivery institutions are well financed and well staffed and can solve the water delivery problem within the existing system and the available means. What is missing is a strong tradition of dialogue and negotiation between the service providers and receivers. This gap cannot be filled through the time-bound intervention of IFI-funded projects because the negotiation process is messy, time-consuming and unpredictable. As OPP's founder Akhter Hameed Khan used to say, the nature of social transformation is in line with the botanist's view of growth, and formal sector professionals have an engineer's vision of growth. If an engineer's approach is taken, it will not succeed.

Learning from the failure of financial patronage

Success, thy name is humility

The cardinal principle for effective provision of drinking water is "Water management is water measurement".[12] No financial or technical assistance package can be designed without having documentation of the population to be served and the measurement of water supplied and appropriate tariff to be charged. IFIs and governments do not possess this knowledge in most of the cases. The solution as pointed out by Rahman and her OPP colleagues lies in (1) mapping and documenting the existing system; (2) managing quota and metered supply to settlements, thus ensuring equitable provision of water; (3) plugging leaks in water mains (both technical leaks and theft); (4) constructing new mains; and (5) accepting the fact that people do finance, construct and maintain lane and secondary water lines (Rahman: 2004: 13). Reform, as noted by Hasan in 1996, cannot be gifted from above, either by a military regime or an IFI (Hasan: 1996: 1). Agency to change unfolds at the individual level. At an individual level, households and industries need to use water more efficiently and judiciously, which again depends on water metering, collection of dues and rationalization of tariff (UNDP: 2017: 2). Responding to the need for systemic reform, one may need to go beyond metering and paying bills. Only citizen participation can lead to systemic reform. That is why Rahman put her life at stake to put an end to the theft of water in Karachi. Her foes thought differently. She was killed on March 13, 2013, for her work in exposing the tanker mafia in Karachi. Such commitment cannot be purchased with loans and grants. For this to happen, citizen institutions need to be strengthened and supported for the long haul. It is in meeting this need that IFIs fail. Keeping in view OPP's extremely effective role in helping improve the water supply situation in Karachi, I will try to explain the difference between the IFI's project approach and OPP's process approach and indicate where IFIs need to change if they want to make the difference. This needs an approach based on humility. As eloquently stated by Ernesto Sirolli, "If you want to help, shut up and listen".

Institutional issues: engagement and partnership

Experience of numerous institutions, both government and non-government, in working with low-income communities has revealed four principles that produce desirable changes: (1) build on what exists – improve and upgrade, don't discard and replace the existing infrastructure built by communities; (2) adapt technology to social condition – meaning determination of the boundaries of the unit for provision of services, and establishing the specifications of design and making cost estimates in accordance with local practices; (3) follow the principles of incremental financing; (4) share components – meaning communities finance, build and manage infrastructure at household and lane level as internal component of development and government finances, and build and operate infrastructure trunk lines and disposal units as external components of development (Hasan: 2009: 3). It is important to note here that improving service provision based on the aforementioned principles calls for a process which is based on an open-ended time framework, supported by availability of funds released in small amounts over a long period of time. It clearly distinguishes the role to be played by different partners and facilitates trust building and cooperation between people, professionals and politicians based on documentation and analysis of local knowledge. People under this framework identify priorities, professionals identify appropriate technical options and politicians allocate the budget. Overstepping these boundaries can cause serious management problems. The financial mechanism needed to support such a programme runs into conflict with IFI-funded and formal sector–supported projects.

Compulsions of IFI-funded projects require them to be time bound, close ended and template based. As a matter of practice, IFI-funded projects have complete disregard for community assets and local practices, and refrain from the preparatory work to build trust between the partners. A major compulsion of IFI-funded projects is to spend large chunks of money in a narrow time frame through quick burners like foreign trips of government officials for 'capacity building', purchase of equipment, vehicles and air conditioners, who measure their 'result-based' success by the amount of project money spent. Spending small amounts of money over a long period of time to build a local body of knowledge and facilitate incremental development increases the administrative or overhead cost of the grants and loans. It is considered inefficient because it takes a long time to complete. It is a different matter that the time taken to successfully complete a participatory, slow-paced project is much less than the time taken by many failed 'professionally managed' projects.

Control over the project funding and delivery mechanisms is the second major compulsion and limitation of the IFI-funded projects. IFIs and development assistance agencies don't have the mandate to use their funds for one-time investment in building local endowments. Creation of such endowments can help projects continue their work in line with local pace over a long period of time, reduce the administrative cost of project management, overcome the limitations of a narrow time frame and undertake a process-based instead of a template-based approach. This kind of financing does not fit in with time-bound, results-based management and a top-down system of financial accountability. It also shows a major trust gap between the IFIs and CSOs engaged in building human and social capital at the local level.

Process-based technical assistance

It would be extremely relevant here to narrate in some detail the working methodology of OPP's founder Dr. Akhter Hameed Khan to clarify the difference between template-based and process-based support mechanisms. In 1980 Dr. Khan was approached by BCCI's CEO Agha Hasan Abidi (1922–1995) to lead a poverty alleviation programme in Karachi's biggest informal settlement, Orangi. Abidi wanted Khan to establish a world-class college to lift the residents of Orangi above the poverty level. Dr. Khan asked Abidi, "how do you know what will help poor rise above the poverty line? Have you talked to them?" Abidi said no. Khan said, "I need to talk to the poor. I would trust their judgement rather than yours or mine". Abidi agreed. But Khan put on the table some other conditions to be accepted if Abidi wanted him to work: (1) an endowment fund would be created to generate a continuing stream of revenue for his project to cover its operational cost and ensure its sustenance, freeing him from the need to keep looking for funds every now and then; (2) Dr. Khan would regularly report on every penny received and spent and the progress of his work; and (3) he would have complete managerial independence and there would be no interference in his work. This gave Dr. Khan the space to talk to people, to identify their priorities and design a programme in accordance with the local conditions. After six months of consultation, he came up with the finding that improved sanitation was the most effective way to alleviate poverty in Orangi.

Dr. Khan's real test started after he presented his ideas to the community and asked them to lay sewerage lines in their lanes on a self-help basis. Everyone in Orangi would agree with him but ask him to lay the first line with his funding. He refused to oblige. It took him another six months before a sewerage line was laid in the first lane. Others soon followed, and as Dr. Khan had predicted there was a snowball effect. People laid sewerage lines in 7,000 lanes on a self-help basis. In his talk at NRSP he described the logic behind the community engagement:

> We could not have done this work ourselves. When they (residents of Orangi) saw an outsider, they would ask him to do it for them. If I went to them and said, "you people make the lane" they would ask, "what are you here for? What do you have?" Their first question was always, "what you have come to give us". However, they could not ask their own man [the activist] what he had come to give them.[13]

This statement clearly shows the strategic importance of giving the lead role to the community rather than the external expert for sustainable poverty alleviation. On another occasion, Dr. Khan said "Bangladesh had received, by 1978, an amount four times the foreign investment made in Germany under the Marshall Plan but did not produce any results". "Progress is never achieved with money but with the dedication and hard work of the workers. When community stands up, idealist emerges, who sacrifice, and their hard work brings progress and prosperity in the community".[14]

In tandem with Dr. Khan's process-based work in Orangi, a UN Human Settlements (UNHS) project following the template-based approach was also in operation. Although the UNHS project believed in the superiority of the template-based approach, it wound up after laying lines in 36 lanes. The funds and project time frame came to end. Both Dr. Khan's OPP and the UNCHS project had received equal amount of budget. The UNCHS project director rejected Dr. Khan's 'unprofessional' and 'unpredictable' approach for not being 'results based'. A close look at the OPP methodology shows that with the UN expertise, budget and time frame, the dream of laying sewerage line in 7,000 lanes would have never seen the light of the day. OPP first prepared a map of the entire settlement of 1 million inhabitants and carried out a level survey to lay the sewerage line so that the flow of disposed wastewater was not obstructed. Dr. Khan started a series of dialogues with local communities to present the data-based case to convince them that (1) lack of sanitation facilities is the major cause of their poverty and (2) with a much smaller amount of investment by each family compared to their annual treatment expenses and lost work days, they can lay a sewerage line in their lane.

Community members were told that if they want to lay a line, they would need to submit a demand notice at OPP office. Submission of their request would mean that (1) they are willing to finance, build and maintain a sewerage line in their lane; (2) they will select a lane manager to interact with them and OPP and lead the whole lane-based project; (3) they will raise money for work in their lane, purchase materials from the market and maintain an account of receipts and expenses; (4) they will hire and supervise the mason for laying sewerage lines and building manholes. In return, OPP will (1) provide them a map and mark the line level in the lane; (2) train the mason in laying the line and use of shuttering technology to build manholes; and (3) train the lane members in technical oversight of the project. It was very labour-intensive and painstaking work. This way of working is called incremental development, where you take one step at a time and end up covering a distance of a thousand miles. It ended in making communities full owners of the project. This approach appears to be slow, small scale and not well directed compared to well-planned and executed projects, but it works in a faster, more frugal and effective way. Such technical assistance would not have been possible with IFI-based costing and time frame, and this was clearly visible in comparison to the UNHS project. The choice is between 'slow' and incrementally financed success and fast, elegant and expensive failure. The process-based approach pleaded by Dr. Khan is summarised in Box 2.1 below.

Box 2.1   How frugal science works: a lesson from Akhter Hameed Khan

"It would be very wrong if I boast that I did this or that. I merely observed a lot of things then founded supporting institutions. But the people did the work themselves."

1. When you contact a community, they are aware of:

- The problems
- They also have ideas for solutions.

1.1 There are two types of solutions:

- Dreams
- Possible within the means.

1.2 Solution within the means reflects community's resource endowment

When you offer solutions within the means, the community's hesitation means anarchy not rejection.

1.3 To build trust with the community:

- Identify early adopters
- When early adopters succeed, there will be a snowball effect.

2. A sustainable solution would be based on:

- Appropriate organization
- Entrepreneur's vision.

2.1 Appropriate organization

Appropriate organization could be any of the following:

- Village organization (VO)
- Activists/early adopters
- Special interest groups.

3. **Whether a function can be privatized or collectivized would determine the nature of organization to accomplish that task.**

3.1 Functions of VO are only:

- Building infrastructure
- Co-op banking.

3.2 Whereas ownership and entrepreneurship cannot be collectivized. They are better done individually

4. **Characteristics of the appropriate organization would be:**

- Decentralization
- Delegation of authority.

5. **The role of the organization would be provision of:**

- Social guidance
- Technical guidance
- Credit.

6. **For sustainability, an organization should aim at:**

- Upgradation not innovation
- It needs to provide entrepreneur's vision
- Follows the principle of incremental development
- Find a solution based on: reduced cost – subsidize research and extension research and extension (our subsidy) and contribution.

The story narrates the strengths and limitations of both the approaches, formal and informal. There are two options: (1) establish a Trust like OPP that can overcome the constraints of project-based funding or (2) continue repeating the failed projects, adding to national indebtedness, poverty of local communities and 'poverty alleviation' of elites engaged in delivering values through the top-heavy value chain. Local Trusts defined the fabric of social life in what now constitutes Pakistan across all the communities of faith, ethnic groups and geographic regions, before the British occupation. This space needs to be re-appropriated by local communities. IFI needs to evaluate if their 'technical assistance' can be wedded to this institutional norm or not. This requires a major policy debate.

Notes

1 This concept was introduced in Pakistan by the Orangi Pilot Project (OPP). The component-sharing approach is based on the concept that local development consists of two components, internal and external. The internal component consists of physical infrastructure at the home and lane level, and the external component includes infrastructure above the neighbourhood level. A sound partnership is formed between the community and government if community takes responsibility for financing and building infrastructure and managing services at the internal level and government does the same at the external level.

2 See for example the recommendation by Sayeed, Husain and Raza (2016). The report specifically recommends "For any formalization to occur in Karachi's land and transport sectors, the local government – with its own 'funds, functions, and functionaries' – must be empowered to develop and oversee the implementation of an inclusive master plan."

3 Computed at the rupee dollar exchange rate of 105.128536 prevalent in September 2013.

4 During the course of discussion in this book, I shall use the term IFIs to denote the common approach of formal sector players in the development assistance industry, including UN agencies, bilateral donors, and national and international NGOs working as contractors of donors and IFIs. The name of a specific agency will be used when a project or point of view of a specific partner is being discussed.

5 Patronage systems are based on the zero-sum thinking, meaning one player's gain is another player's loss (e.g. the community's gain is the government's loss). Nonzero thinking means that one player's gain is every player's gain. If the community gains development funds and services, it pays for the services as well and creates a public good that benefits everyone.

6 These figures are based on the household survey of 4,500 houses located in nine out of 18 town governments under City District Government Karachi by WSP (WSP: 2010: 23).

7 Safina Siddiqi was an office bearer of Karachi Administration Women Welfare Society (KAWWS) in 1990s. After failing to have her complaint redressed by every government agency, she approached the court where she found a supporter of her just cause. The court allowed KAWWS residents to stop paying taxes until the district government redressed their complaint. This case has been documented in Hasan (1998).

8 During my interaction with municipal committees in Lodhran and Dera Ismail Khan, I found out that their record of property and assessment of its value for charging property tax was 20–30 years old.

9 I interacted closely with numerous local government authorities as coordinators of UNDP's small grants programme, UN's Civil Society Advisor and development expert with Tehsil (Town) Governments of Multan and Jaranwala, Municipal Committee of Hyderabad, Lodhran and Dera Ismail Khan, Capital Development Authority (CDA) and provincial Planning and Development Departments of Sindh and Punjab. None of the local governments had the updated city map. In the case of Municipal Committee Lodhran, there was *no map* at all. At this time, Survey of Pakistan (SOP) was compiling maps of all the towns of Pakistan. I requested a map of Lodhran from SOP through my colleague who was head of the governance unit in UNDP and administering grant fund to SOP (apparently for improvement of governance). SOP declined to share the map on the ground that these maps are 'confidential'. When the director general of SOP was told that these maps are available with American, French, British, Russian and even Indian vendors, it did not help. The point is that when one government department does not share information with another department and even with donor, how can donor funding help in doing any meaningful planning?

10 Pakistan's local government system consists of three tiers. The lowest tier is Union Council (UC); next higher tiers are Tehsil (sub-division) or Town Municipal Administration (TMA) and district government. Anjuman Samaji Behbood – led by one-person pro bono technical assistance mission Nazir Ahmad Wattoo – worked in the low-income communities of UC 62 Lahore and Bhalwal Town to institutionalize a metre-based billing and service delivery system. In the first case it started from scratch and demonstrated a model of participatory water management based on user fee payment; in the second case it revived a dysfunctional water supply system following the same principle. These experiments endorsed the views upheld by informal community's support organizations in Karachi. Work of many model CBOs in Punjab has shown the effectiveness of this principle as well.

11 Details of other technical assistance projects are not given here because ADB, World Bank and in UNDP funded Water and Sanitation Programme (WSP) and Japanese-assisted programmes are similar. See for example Water and Sanitation Programme (2010) and annual AJCE Newsletter, April 2009, page 1.

12 Anjuman Samaji Behbood (ASB) is a community-based organization in Pakistan. ASB's model for improving water supply without donor and government funding called Changa Pani has been replicated in in Bhalwal Town, and now the Punjab government is trying to replicate it in the other 56 municipalities of Punjab.

13 Dr. Akhter Hameed Khan "Looking for the Man" 1998, p. 2 NRSP.

14 UNDP, Report on 3rd LIFE Grantee NGOs Workshop, 1998, Faisalabad, Islamabad.

References

ADB, 2010, *Pakistan: Mega City Development Project*, Completion Report, Project Number: 38405, Loan Number: 2229-PAK.

AFP, 2013, "Karachi's Black Economy Generates Rs830 Million Daily", *Geo News*, viewed from www.geo.tv/latest/63555karachisblackeconomygeneratesrs830milliondaily.

Ahmed, N., 2010, "From Development Authorities to Democratic Institutions: Studies in Planning and Management Transition in the Karachi Metropolitan Region Commonwealth", *Journal of Local Governance*, No. 7. Online viewed from http://epress.lib.uts.edu.au/ojs/index.php/cjlg.

Ahmed, N. and M. Sohail, 2003, Alternate Water Supply Arrangements in Peri-urban Localities: Awami (people's) Tanks in Orangi Township, Karachi. *Environment and Urbanization*, Vol. 15, No. 2.

Askari, S., 2015, *Studies on Karachi*. Papers Presented at the Karachi Conference 2013, Cambridge Scholar Publishing.

Daily Times, 2016, "7m Gallon Water being Sold 'illegally' by Hydrant Mafia", viewed from http://dailytimes.com.pk/sindh/23May16/7mgallonwaterbeingsoldillegallybyhydrantmafia.

Dawn, November 10, 2016, "Bribes and Shortages: Karachi's Burgeoning Water Mafia Pakistan", viewed from https://www.dawn.com/news/1295479

Dawn, October 4, 2015, "The 'Water Mafias' that Suck Karachi Dry", viewed from https://www.dawn.com/news/1210853

Dawn, March 22, 2011, "Water Woes in Karachi Pakistan", viewed from www.dawn.com/news/print/615047.

Dawn February 22, 2007, "KWSB Revenue Increases by 200 pc", viewed from https://www.dawn.com/news/234032/karachi-kwsb-revenue-increases-by-200pc

De Soto, H., 2000, *The Mystery of Capital: Why Capitalism Triumphs in the West and Fails Everywhere Else*, Basic Books, New York.

Haider, I., 2007, "Reforms in KW&SB – An Overview for Civil Society", viewed from http://arifhasan.org/karachi/reforms-in-kwsb-an-overview-for-civil-society.

Hasan, A., 2017, "Karachi Diagnostic", *Dawn*, January 8, 2017, viewed from http://arifhasan.org/articles/karachidiagnostic.

Hasan, A., 2012, "The Anti-Poor Bias in Planning and Policy18 November, 2012", viewed from http://arifhasan.org/articles/the-anti-poor-bias-in-planning-and-policy.

Hasan, A., 2009, *The Water and Sanitation Challenge: The Conflict Between Reality and Planning Paradigms*, Outline of a paper presented at the Aga Khan University's 13th National Health Sciences Research Symposium on "Impact of Water and Sanitation on Health", held on October 27–28, 2009 in Karachi, http://arifhasan.org/seminars/thewaterandsanitationchallengetheconflictbetweenrealityandplanningparadigms.

Hasan, A., 2006, "IFI Loans and the Failure of Urban Development", *Dawn*, viewed from http://arifhasan.org/articles/ifiloansandthefailureofurbandevelopment.

Hasan, A., 2005b, "Some Water and Sanitation Related Issues: Initial Thoughts", *Archi Times*, viewed from http://arifhasan.org/articles/somewaterandsanitationrelatedissues initialthoughts.

Hasan, A., 2005a, The Orangi Pilot Project Research and Training Institute's Mapping Process and its Repercussions, March 14,2005, viewed from http://arifhasan.org/wp-content/uploads/2012/07/The_OPP-RTIs_Mapping_Process.pdf.

Hasan, A., 1998, *Community Initiatives: Four Case Studies from Karachi*, City Press, Karachi.

Hasan, A., 1996, "Hijacking the Process", *Herald*, viewed from http://arifhasan.org/articles/hijackingtheprocess.

Hasan, A., Sheela Patel and David Satterthwaite, 2005, "How to Meet the Millennium Development Goals (MDGS) in Urban Areas", *Editorial Environment & Urbanization*, Vol. 17, No. 1.

Hordijk, M., 2005, "Participatory Governance in Peru: Exercising Citizenship", *Environment & Urbanization*, Vol. 17, No. 1.

Rahman, P., 2004, *Katchi Abadis of Karachi: A Survey of 334 Katchi Abadis*, Orangi Pilot Project-Research and (OPP-RTI) Training Institute.

Rahman, P., 2008, "Water Supply in Karachi: Situation/Issues, Priority Issues and Solutions", *Orangi Pilot Project*, Karachi.

Rahman, P., 2013, "We Are All Ninja Turtles of Mapping", Perween Rahman's talk at ACHR. Online viewed from www.achr.net/upload/downloads/file_13122013114415.pdf.

Sayeed, Asad, Khurram Husain and Syed Salim Raza, 2016, *Informality in Karachi's Land, Manufacturing, and Transport Sectors: Implications for Stability*, USIP.

Sharif, A., 2008, "Defaulting Government Agencies May Lose KWSB Connections", *Dawn-18*, viewed on February 13, 2008.

Stevenson, Bryan, 2012, "We Need to Talk about an Injustice", *TED Talk*.

UNDP in Pakistan, 2017, "Water Security: Pakistan's Critical Development Challenge", *Development Advocate Pakistan*, Vol. 3, No. 4.

UNDP Pakistan, 2015, *Annual Report*, Islamabad.

Water Aid, 2007, *The Advocacy Sourcebook*, Water Aid, London.

Water Aid, 2009, *Case Study: Pakistan, Civil Society Organisation Involvement in Urban Water Sector Reform in Pakistan*, viewed from www.wateraid.org/urbanreform.

Water and Sanitation Programme (WSP), 2005, *Shehr Ki Dunya Workshop Report on Managing Karachi's Water Supply and Sanitation Services.*

Water and Sanitation Programme (WSP), 2010, *Water and Sewerage Services in Karachi. Citizen Report Card: Sustainable Service Delivery Improvements.*

Zulfiqar, Q., 2011, *The Express Tribune*, Sindh Karachi suo motu, Verdict out Published, viewed on October 6, 2011, Pakistan.

3 Informal settlements, land mafia and failure of government policy

Background and context

Most of the areas which now constitute Pakistan are part of the Indus Valley. There is a legacy of building towns with highly developed systems to meet the needs of local population based on local climate, building materials and lifestyle. This tradition further evolved during the rule of various dynasties up until the Mogul period. During British rule, new urban centres and hill stations were developed blending tradition with the needs of British administration. After independence Pakistan faced the key issue of expanding urban population due to massive in-migration from India and subsequent rural-urban migration. In the early years, Pakistan had to find a quick housing solution for Muslim refugees who crossed the border from India after partition in 1947. Karachi's population almost doubled in the course of a few years. Pakistan's port city Karachi emerged as a new commercial and industrial centre as well as the capital. A new class of urban workers and low-income service providers subsequently became part of the city and was in need of shelter. There was no experience of town planning among Pakistan's professional class. The Ford Foundation and its town planning experts and a Swedish firm, Merz Rendel Vatten, offered their services to the Government of Pakistan. Due to intensive lobbying by the Ford Foundation, Greek architect and town planner Doxiadis was picked as the adviser for proposing a city development plan for Karachi and subsequently for Pakistan's new capital Islamabad.

Both in the case of Karachi and Islamabad, Doxiadis could not propose a plan that would effectively address the issue of providing shelter to the poor. This is the dilemma of the entire formal sector. At present Pakistan has a housing backlog of almost 12 million units. Housing units developed are way beyond the capacity of low-income groups, who constitute a majority of the population. According to a survey by the State Bank of Pakistan, the affordable housing price to income ratio is 20:1 in Pakistan compared to a global average of 5:1. A home mortgage market that could partially help overcome this problem has not developed in Pakistan. Commercial banks' housing loans cater to high-income groups, and public-sector House Building Finance Corporation (HBFC) does not serve low-income groups either (Baig: 2017). Government policies and budgetary support mechanisms have not achieved much success in solving the problem. The failure of housing policy is partly caused by following the solutions out of sync with the local reality. Government policies and plans while upholding the right to shelter have not been able to deliver. A brief review of practices that originated in Karachi and were subsequently implemented in Islamabad will help clarify this point.

Planning for providing shelter to the poor

Pakistan's experimentation for providing shelter to the poor started in Karachi soon after independence. Karachi's population expanded on an exponential scale after 1947. Between 1941 and 1961 it increased by almost 200 percent. Local government did not have the capacity to cope with such a massive increase in population. While the 1960s witnessed rapid industrialization, the 1970s ushered in expansion of the informal sector due to nationalization of industry and disintegration of the textile sector. Massive export of drugs and import of arms during the Afghan War years in the 1980s and 1990s further expanded the informal economy. It was followed by privatization and deregulation in subsequent decades. As a result, informal settlements now meet the housing needs of 50 percent of Karachi's population (Sayeed et al.: 2016: 5). During the past 70 years, five Master Plans were prepared for the city and failed to provide formal, low-cost housing to the poor (Sayeed et al.: 2016: 3). In 1952, Merz Rendel Vatten presented the idea for a new capital area in Karachi to separate the administrative zone from highly congested dwellings of the low-income population. However, even this solution did not seem to work well. It was eventually decided to move the capital to a new location. A new location offered the opportunity to build the city from scratch, without any baggage of the past or any need to undo the unplanned construction activity. Generous funding and world-class technical advice was provided for starting this dream venture. This perfect attempt at building a modern city also did not succeed in providing shelter to the poor. A constitutional provision in 1972 to provide shelter for the poor did not help either.

Islamabad, the case of starting urban planning from scratch

In 1960, the government of General Mohammad Ayub Khan decided to establish the new federal capital at the foothills of the Margalla Mountains. The new city was named Islamabad and the Greek firm of Doxiadis prepared a Master Plan for the new capital. "Doxiadis had been at the forefront of a new breed of entrepreneurs who turned strategic development funding by Western governments into big business" (Daechsel: 2013: 88). Constantinos A. Doxiadis, who had earlier planned Asia's largest refugee settlement in Korangi with funding by the Ford Foundation, was tasked to prepare the Master Plan for Islamabad. Doxiadis's plan for settling the poor in Karachi proposed to create separate housing units for the poor, far away from the government officers' residences and envisaged transporting workers back and forth through public transport. The distance of this proposed housing scheme from the place of work made transportation so expensive that workers lost interest in availing themselves of this facility. Despite this failure Doxiadis was given the contract to design Pakistan's new federal capital (Daechsel: 2013: 92).

Islamabad provided an ideal location for experimentation of Doxiadis's solution of separating the poor from the rich for meeting their shelter needs. It suited Pakistan's military and civil bureaucracy as well. Islamabad was home to a few sparsely populated villages that could easily be relocated to build the new capital without any baggage of the past, burden of the poor and 'corrupting' influence of people from other walks of life. It created an opportunity to build the new capital like the colonial Hill Station in South Asia, "an idealized mountain 'village' from where the most striking features of urbanity had been deliberately removed" (Daechsel: 2013: 87). Some experts have defended Doxiadis's failure in planning housing for the poor based on the view that his plan to settle the builders of the city was neglected by the Pakistani decision makers. He could not develop a plan for the builders and service providers as his contract was prematurely cancelled after he completed plans for a few sectors (Mahsud: 2007). This argument does not carry much weight in view of the plan for sheltering the poor he had developed for Korangi. Although Doxiadis's vision included a dynamically growing city in an inclusive way, in reality his plan included little in the way of accommodating the poor. His concept of the poor also did not go beyond the low-rank government employees and domestic servants.

Apparently Doxiadis's view of a modern, well-planned city attempted to provide space to all groups and classes of people, but his layout plan for the city was not much different from the civil military officer elites' view of a centre of government, keeping away the civilian population to pre-empt corruption. He wanted the city to have no connection with the past and government wanted it to be a modern version of the colonial Hill Station, replicating the division based on cantonments, civil lines and residential areas. Both had one vision in common: opposition to unplanned, irregular mixing of people overruling rank and status of the residents. If at all, his concept of inclusion applied only to lower-rank government employees and domestic servants. There were separate sectors allocated to lower-rank government employees, and within government officers' residences there were servant quarters for domestic servants. His plan was silent about the service providers of the city: construction workers, milkmen, sweepers, fruit and vegetable vendors and numerous categories of workers in the private sectors. For the service providers, he had mentioned the idea of 'follow on' without providing any concrete proposal or having separate and distant enclaves implying high transportation costs not affordable for most of them. This concept of inclusion had no connection with the ground reality.

The view he had of settling the poor was no different than his plan to settle the poor in Karachi. More than the intent or resources, the problem of informal settlement has arisen due to lack of understanding the problem of housing for the poor and their relationship with the city. This lack of understanding created a bias against finding an amicable solution for the shelter of poor which has continued during the past six decades irrespective of regime and policy changes. It is important to note here that while successive planners and administrators could not find enough land for thousands of low-income service providers, they have been able to accommodate middle and business classes by allowing construction of high-rise buildings, expansion of commercial areas, development of entertainment parks, provision of land for world-class shopping malls and altering land use plans by making provision for Metrobus services and flyovers. At the same time, they have continued demolishing the informal settlements under various pretexts – being an eyesore for the city, a security threat to the neighbouring sector, illegal encroachment of land, lack of availability of land and so forth.

Doxiadis's concept of urban planning was based on two key concepts of dynapolis (being dynamic and having the capacity for endless growth from one end to the other as a metropolis) and social mixing (Daechsel: 2013: 88; Mahsud: 2008). The city plan allocated specific sectors for residents in each income category, provided areas for interaction between various classes of residents and room for the city to grow and continuously accommodate increasing population with the passage of time (Kreutzmann: 2013: 140). The Capital Development Authority (CDA) was constituted under the CDA Ordinance of 1960 and tasked with planning and developing the federal capital in accordance with the approved Master Plan. The city did not allocate any areas for low-income groups other than the government employees. Servant quarters were made part of official houses to accommodate domestic help. For the city's construction workers and service providers, Doxiadis planned to provide thorough follow-up, but nothing was provided in the Master Plan. The issue cannot be explained in terms of weak governance because CDA had all the powers as a planning authority that other municipalities in Pakistan either lack or do not exercise (Ahmad and Anjum: 2012). Due to systematic neglect of the poor, informal settlements emerged to cater to the needs of service providers as well as low-income employees, and migrants coming to the capital in search of job opportunities. Many of these low-income residents work for CDA. CDA solved the problem of shelter by ignoring informal settlements or periodically evicting them in the name of compliance with the Master Plan.

During the planning stage the question of merging or separating Rawalpindi in the Master Plan was seriously discussed. Rawalpindi's inclusion was rejected on the pretext of keeping the 'corrupting' influence of non-bureaucratic classes. The Master Plan for Islamabad called for Islamabad and Rawalpindi to expand indefinitely in parallel spaces. Islamabad was to be divided by a green belt and a transportation passage. Starting with grand administrative complexes on the one end of the city, residential sectors were allocated a numbered and lettered grid of 1.25 square mile sectors extending down a gentle slope. Each sector was divided into a nested hierarchy of 'communities', each with a school, a market and a mosque commensurate with its population. In order to establish a transparent relationship between social position and physical dwelling, planners prescribed a hierarchy of state housing sizes and designs corresponding to residents' salary level and position in the government bureaucracy. Doxiadis praised Islamabad's grid because it 'can develop dynamically, unhindered into the future, into space and time' (Doxiadis: 1965: 26; Hull: 2009: 443). Mahsud has drawn attention to the underlying contradiction in Doxiadis's planning concept. While the plan provides for limitless growth of the city with development of new sectors along the planned grid, it leaves no room for the expansion of developed sectors in existence. This simultaneous existence of dynamic expansion and planned stagnation has created serious constraints for the city to accommodate newcomers, especially in the low-income sectors (Mahsud: 2007). The sectors built on the concept of social segregation and stagnation kept in place the hierarchical character of the city and led to housing shortages for government employees as well (Kreutzmann: 2013). The problem of housing shortages in extremely well-planned sectors of Islamabad has led to many malpractices and due to the enormity of the problem CDA is completely helpless in eliminating many forms of illegal occupation and renting out of government employees' quarters, land-grabbing and encroachments and violation of zone requirements (Adeel: 2010). Deficiencies of planning and management are always explained in terms of deficiency of resources and weaken any effort to rethink the city's problems from the point of view of the poor.

Informal settlements

Poor service providers of Islamabad responded to the official plan's neglect of the poor in the name of steering away from the 'corrupting' indigenous influence of the low-income population by undertaking unplanned appropriation of land. Unplanned use of land has taken place in two ways. First, unplanned settlements have sprung up at various spots in Islamabad in tandem with well-planned sectors (Malik: 2017). Second, neglected residents have undertaken quiet encroachment of plots and small pieces of land in connivance with the CDA employees. Individual encroachers manage misplacement or 'loss' of a file or take out-of-turn possession of government flats by using their connections inside the government, or by peddling influence, resorting to threats or greasing the palms of CDA employees to lay claim to property (Hull: 2009: 445). Another practice is for occupants to rent out a part of the government-allotted houses to private individuals. This practice is very common, and many government employees rent out a room to single government employees, job seekers or newcomers to the city. Encroachments have also taken place by construction of numerous unplanned mosques on illegally occupied piece of land.[1] The Federal Shariat Court (FSC) regularized all such mosques by its decision in 1984, and this practice continues without impunity (Hull: 2009: 462, Beacco: 2018). In some cases, rural migrant workers have started sleeping in corridors of the sector markets.[2]

There were three types of unapproved settlements: those occupied by the city's low-income service providers; those inhabited by native villagers whose villages fell under the boundaries of Islamabad's Master Plan; and those middle-class encroachers who purchased the land and constructed houses and business enterprises in violation of CDA's rules and regulations. CDA failed in reversing the course taken by all three forms of settlements. While CDA grudgingly accepted the latter two forms of violation of the Master Plan, it did not accept the low-income informal settlements as a matter of policy. CDA tried various ad hoc solutions to deal with the growth of informal settlements: eviction, relocation to a formal settlement and rehabilitation. These policies suffered from the limitations of a formal sector approach and did not provide any lasting solution (Beacco: 2018). It is very informative to look at some of the initiatives taken by CDA to deal with the residents of these settlements.

Evictions – tyranny of ideology

First and most common solution employed by CDA was to evict or demolish an existing informal settlement.[3] This is a good example of blaming and penalizing the victims. Service providers who built Islamabad and run it on daily basis are penalized for the failures of planners, administrators and city managers. Starting in 1979 the state of Pakistan upheld the right of homeless people to shelter. CDA had ample resources to accommodate these homeless people in an amicable way, but the tyranny of our ruling elite's ideology has pre-empted any such solution. The ideology of eviction is simple: it is poor people's responsibility to fill the government coffers through payment of indirect taxes to the tune of 60 percent of tax collection, but it is their own responsibility to find shelter even if they cannot afford it. Many of these settlements have been demolished and re-emerged several times. They are scattered all over Islamabad as shown in Figure 3.1 below. On the question of shelter, in Islamabad *all the people are equal, but some people are more equal than others.*[4]

According to one survey, 33 informal settlements known as katchi abadis (KAs) exist in Islamabad by now, but the number has changed with the passage of time. The pretext used for eviction varies from being an eyesore to posing a threat to the nearby sector to violation of CDA rules. These moves have never succeeded in solving any problem. Residents of KAs always find an alternative spot to build their huts again. They grudgingly accept the financial losses after eviction and life comes back to normal when they move to a new location. In some cases, local advocacy groups, elected representatives, political leaders and even cabinet ministers have raised their voices against eviction, and CDA had to withdraw and discontinue its eviction work.[5] Many of the residents living in these settlements are CDA employees themselves and find ways to pre-empt eviction. CDA has also accepted this reality with discomfort. Some CDA administrators therefore tried alternative approaches to settle the issue on a long-term basis. The National Shelter Policy announced by President Pervez Musharraf made it mandatory for

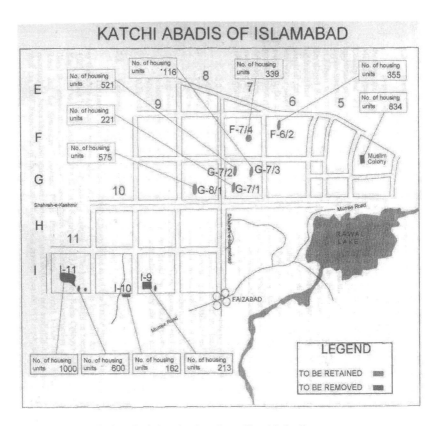

Figure 3.1 Map of Islamabad showing location of katchi abadis

Source: Husain 2002

municipal administrations not to evict occupants of informal settlements without providing an alternative spot, but CDA's eviction operations continued in complete disregard of this policy. To counter these illegal operations a national coalition of activists for housing rights, All Pakistan Katchi Abadi Alliance (AAPKA), has petitioned in the court to take the court's order against eviction in some informal settlements. Continued failure of attempts at eviction strengthened the case for finding a permanent solution.

Rehabilitation

Another alternative tried by CDA was to rehabilitate the informal settlements. Rehabilitation meant clearing an existing informal settlement, converting it into plots of equal size on a grid pattern, providing access to all the basic services, giving rights to evicted families to purchase plots on payment of a specific amount as down payment and transferring the purchase deed to the payee. CDA tried this approach in an informal settlement in a sub-sector of F-6/2 (known as 100 Quarters), a sub-sector of G-8 renamed as Christian Colony and another sub-sector of F-10 known as Shalimar Colony. Under this approach, the land cleared from squatters was divided into smaller plots of 3 *marla* size on a grid-iron pattern like other sub-sectors with larger-size plots. All the services and basic amenities available to the rest of the sectors were provided as well. This solution ran into two problems. First, due to the larger size of plots than the existing houses of the squatters, all residents could not receive plots in rehabili-tated schemes. Second, the residents were provided plots at subsidized rates, and due to regularization of their area the commercial value of the plots shot up. It tempted most of the allottees to sell their plots and move out of the reha-bilitated colony. In both cases, the target residents vacated the existing informal settlements and created a new informal settlement. In the case of 100 Quarters in F-6/2, CDA allowed an NGO formed by a political leader to construct low-cost houses in 190 plots of 20 × 30 feet size for informal settlement dwellers. The agreement was cancelled due to charges of corruption against the NGO and rehabilitation work was stopped. The NGO had received Rs. 55,000 in advance for allotment of the plots to the squatters, and after receiving the payments they started selling the plots to outsiders (Husain: 2002). This did not help CDA in solving the problem of shelter for the poor.

Relocation

Relocation provided the opportunity to evicted residents to receive the title for a plot planned like a rehabilitated colony away from their current location. In the early 1990s, CDA removed a squatter settlement located adjacent to the Fatima Jinnah Park in F-9 sector and agreed to provide small plots (20 × 40 feet) free of cost to the evicted families in in Alipur Farash, a settlement located 10 kilometers away from the city. Occupants were told that they would not be disturbed for 15 years. Due to high commuting costs, the occupants changed their decision and more than 30 percent of allottees came back to a new katchi abadi after selling their plots (Husain: 2002). KA residents of Muslim Colony were also persuaded to move to this site. Muslim Colony was perhaps the first and largest informal settlement which housed the construction workers who started building Islama-bad. Since this colony was situated on the back of President House, CDA was determined to get it vacated due to security 'concerns'. It did not work out. Alipur Farash was too far away from their areas of work and it added a considerable amount to their daily expense in the form of transport fare. Muslim Colony was situated next to the shrine of a renowned saint, Bari Imam, and almost all the resi-dents could regularly avail themselves of the opportunity of getting a free meal from the shrine's free kitchen (*langar*). Other efforts to relocate the squatters met a similar fate.

Vacating local villages for sector development

Another group of low-income residents consisted of the native villagers of the areas which were made part of Islamabad. CDA permitted the villagers to continue their stay until the development work started in their areas. CDA was bound to meet all the formalities for land acquisition during the development phase, and one of the conditions for getting possession of the land was to compensate the villagers for the land procured for development. Villagers were willing to vacate the land, but CDA offered them pre-1960 land rates while the prices had hit the ceiling at the end of the millennium. Villagers at other undeveloped locations like Shahzad Town, Humak Town, Rawal Town and Bani Gala fetched much higher prices from the clients compared to CDA's compensation package.[6] CDA's compensation rates could have rendered these villagers homeless because they could not buy a piece of land at these prices anywhere near Islamabad. CDA's efforts to vacate these villages were met with fierce resistance, and at a couple of locations there was exchange of fire between the police and local residents. On both occasions a resident was killed by police fire, the situation turned explosive, the police retreated, and the operation wound up. Very influential local residents also provided moral support to the residents and the Master Plan was put aside. In case of Bani Gala, for example, CDA opposed construction of houses because disposal of liquid waste in Rawal Lake could lethally contaminate the source of drinking water for residents of Islamabad. But Pakistan's icon Dr. A. Q. Khan was among the first ones to build his house there followed by another icon, Imran Khan, and many advocacy NGOs. World Bank supported grant funding entity Pakistan Poverty Alleviation Fund (PPAF) followed suit and set up its office complex in the same area (Moatasim: 2015, 2017).

Violation of Master Plan was nothing new in the case of the higher-ups. Non-farmers including the former chief of army staff, for example, inhabit the entire area allocated for farm houses. You can find international consultants, policy advisors, poverty experts, good governance champions and media icons comfortably nested in the cosy quiet of the farm houses. Violating the Master Plan for Margalla Hills National Park, a blue-eyed federal minister of President Musharraf, in partnership with some greedy builders, started construction of chalets at Margalla Hills to be sold to the filthy rich at sky-high prices. It took the wife of a Pakistani diplomat to raise hell on this issue and personally approach President Musharraf to take firm action to permanently stop this illegal construction. Zoning requirements were violated also by international donor agencies carrying the banner of good governance, NGOs, private schools and businesses.[7] It was violation of the Master Plan by the poor, who resorted to illegal occupation due to denial of their rightful claim to shelter, that bothered authorities.[8] This reminds one of Akhter Hameed Khan's repeated statement that "Pakistan's problem is moral not economic."[9]

Upgradation – learning from SKAA

While CDA was at loss to find a suitable option to deal with the issue of shelter, another government agency in Karachi, Sindh Katchi Abadi Authority (SKAA), had successfully tested a model for providing shelter to the poor by upgrading and regularizing informal settlement. Tasneem Siddiqi, director general of SKAA, had observed over the years the process followed by land-grabbers in setting up informal settlements. He observed that while plots in the government housing schemes for the poor would lie vacant for years, even decades, informal settlements would immediately draw clients in large numbers to purchase land in these illegal settlements (Hasan: 2002). Low-income buyers would take possession of the plot, move there with their families and start building the houses. Land-grabbers would also arrange for all the basic services, provide construction materials on credit, give technical advice and arrange for basic services in a short span of time. Soon the illegally grabbed area would be humming with activity. Siddiqi's keen observation of the informal settlement development process provided him important clues in designing his own regularization and upgradation process in communities under SKAA.

Tasneem Siddiqi designed an alternative government model for housing the poor based on the land-grabber model during his tenure as head of the Hyderabad Development Authority (HDA). He named it "Khuda Ki Basti" (God's abode) and tested it on a small scale. The model worked out well and was later replicated in Karachi under the auspices of SKAA. In SKAA Tasneem Siddiqi introduced some radical changes in the management practices. First, he moved the field office of SKAA to informal settlements and changed the office hours to after work hours to enable applicants to visit the SKAA office without taking leave from work and spending significant amounts on transportation. He also drastically cut down the paperwork and the steps required for completing the application process. Development of the area and provision of services was undertaken gradually, and land charges were recovered in easy instalments. Lease title to the purchaser was issued after the entire family had moved to the location, house construction had been completed and all the dues had been cleared. This resulted in regularization and upgradation of hundreds of KAs during his tenure. He then decided to introduce his model to other katchi abadi authorities and development agencies across Pakistan.

Tasneem Siddiqi's first challenge was to incorporate the lessons learned from the land-grabbers into the management practices of SKAA. He needed to create a friendly and accessible department following simple and effective procedures and trusting the residents of katchi abadis. SKAA's approval process required provision of a long list of documents (Zaidi: 2001). Numerous verification measures were taken by SKAA employees after receipt of documents to take a decision on the application. Some of the key requirements are given in Box 3.1.

**Box 3.1   Art of doing simple things in difficult way.
Experience of SKAA**

1   Application on prescribed form along with supporting documents mentioned below:

Areas councillor's certificate verifying occupation on the plot.
Verification by the witnesses.
Affidavit stating bona fides of the dweller on the plot.
Copy of the national identity card.
An undertaking expressing responsibilities in case of misstatement regarding ownership of the plot.
Other documents like ration card, electricity bills and so forth supporting continuous occupation of the plot.

2   Scrutiny of papers and documents by concerned department.
3   Checking plot size (dimensions) and land use.
4   Preparation of site plan and calculation of regularizable area on prescribed form.
5   Occupant removes structure falling under planned proposals.
6   Issuing demand note (*challan*) based on approved lease rates.
7   Payment of lease charges by applicant in a scheduled bank.
8   Issue of lease deed to the applicant for affixing stamp.
9   Execution of lease by sub-registrar after payment of stamp duty by the applicant.

(Source: Husain: 2002)

Tasneem Siddiqui noted that the cumbersome leasing procedures listed in Box 3.1 required several visits to different offices. Most katchi abadi dwellers were daily wage earners. They could not afford to waste their time in attending offices for months altogether. He also observed that the element of corruption discouraged poor people in taking leases. Bribery rates ranged between Rs. 400 and Rs. 1,200 per lease, which is a substantial amount for poor people. Finally, he noted that the regularization process was not transparent, that even the lease rates were not known to the people. Another serious problem was that plans were prepared with the conventional top-down approach and were seldom discussed with the people. Tasneem Siddiqui observed that due to the prevalent mode of planning and development, lease rates had become unaffordable for the katchi abadi dwellers. Donor agencies and the government insisted that the upgradation programme should be used as a revolving fund. Previously, when the local councils were using their own funds for the upgradation work, this condition did not exist. But there was no accounting system to see whether lease charges were commensurate with expenditure on development work. In the words of Tasneem Siddiqui:

> Here was the dilemma. The current mode of development/upgradation is that the consulting firms make the plans ignoring all existing development work. They use British or American standards, and over-design the services. Resultantly they make the development work very expensive. Implementation is done through the contractors and is supposed to be supervised by the engineers. This consultant-engineer-contractor combine increases the cost by at least 100 per cent which the community is supposed to pay. Add another 30 percent for the kickbacks.
>
> (Husain: 2002)

Now, the question is why should the people pay for the services (which are generally sub-standard because of lack of supervision) which in most cases they already have? They, therefore, do not come forward to take the lease. The option for the local council is either to give subsidy on this count bearing the loss itself or meet it from the funds received from the donor agencies/federal government or not issue the lease. SKAA's action research indicated that in most of the informal settlements people build most of the physical infrastructure themselves. Keeping this in view, development charges in KAs could be reduced without subsidizing the applicant or borrowing from an IFI. Judicious use of development charges collected from the applicants could enable SKAA to cover its own operational expenses as well, and many KAs could be regularized immediately.

The next step was to tailor SKAA's financial system to the land-grabbers' system of incremental financing. Tasneem Siddiqui found out that the first important factor was the cost of the plot and the upfront payment required to move in. The cost of government plots was so high that no poor families could afford to buy it; the upfront payment was also very high. Poor clients could not borrow from the public or private sector to arrange for even the down payment in the official schemes. Since only middle-class applicants could afford to pay for the land, they would buy plots in the name of the poor and then keep them vacant for a long time for speculation purposes. This explained vacant plots in government schemes and the presence of poor occupants in informal settlements. Part of the reason for the high cost of plots in government schemes was high development cost and sale of plots after full development of the residential schemes. Land-grabbers solved this problem by selling land without development, demanding a low upfront payment, giving possession to the client after the whole family moved in to settle down, undertaking incremental land development and allowing the client to build the house incrementally. This arrangement was in line with the financial reality of the poor. There was no intimidating and complex paperwork. Land-grabbers also provided technical guidance through contractors and arranged credit for incoming families to start building the house. During the process of house building, land-grabbers gradually arranged for all the services required by the occupants: water supply through water tankers, sewerage disposal through soak pits, transport to the place of work and other similar matters. This resulted in high level of occupancy. They ensured prevention of evictions by greasing the palms of government functionaries. In some cases, they would also ask low-level employees in water and power supply departments to issue bills to some residents for payment of services even if no services were approved for the area. Bill recipients would continue paying the bills and later on take these bills to issuing departments to demand provision of services.

Establishment of Katchi Abadi Cell in CDA

In 1995 Tasneem Siddiqui took the lead in organizing a workshop in collaboration with UNDP, CDA and the United Nations Economic and Social Commission for Asia and the Pacific (UNESCAP) to deliberate on "Community Based Low Income Housing for Rawalpindi/Islamabad Metropolitan Area". CDA Chairman Qamar Zaman Chaudhry showed keen interest in learning from SKAA's experience. The workshop was held in December 1995, and CDA officials, residents of katchi abadis and selected NGOs and donors were invited to attend. A thorough exchange of views between the key partners resulted in the recommendation that wherever possible CDA should undertake efforts for regularizing and upgrading katchi abadis at the existing locations in an organic way to cater for the shelter needs of the poor and the downtrodden. As to be explained later, this concept of developing KAs in an organic way on the existing location was a major turning point in finding a long-term solution to housing the poor. Subsequently a Katchi Abadi Cell (KAC) was established in 1998 within the CDA Planning Wing. This cell was formed to coordinate and consolidate ongoing efforts made by CDA for solving the katchi abadi problem and, in particular, to contribute towards monitoring, and rehabilitation and pre-empting further creation of katchi abadis.

KAC was given the task of documenting physical, social and demographic situation of KAs in Islamabad and developing a knowledge management system to help CDA regularize these KAs and transfer lease documents based on sound evidence. The key tasks given to KAC were (1) to carry out housing and demographic survey and document land use pattern of various existing katchi abadis; (2) to digitize data relating to katchi abadis and establish a database and a geographic information system (GIS); (3) to coordinate with UNDP, SKAA and community leaders; (4) to prepare improvement plans of katchi abadis and issue approved plans to all executing agencies; and (5) to monitor illegal construction, coordinate with directorates of municipal administration and enforcement of regulations to control expansion of katchi abadis.

Upgradation work under KAC

Phase I of the project envisaged the implementation of the following activities:

Updating and verifying the physical survey, house numbering and counting, establishing linkages with existing community organizations and leaders, surveying the built up area and transferring of built up area on drawings, preparation of list of residents, finalization of list of residents after displaying and inviting objections in open community meetings, conducting socio-economic survey, preparing a database and Geographic Information System (GIS) and preparing improvement and readjustment plans.

(Husain: 2002)

The activities to be carried out under the project were classified under two types: upgradation and rehabilitation. These activities are reproduced in Box 3.2.

Box 3.2   Upgradation and rehabilitation work undertaken by KAC

Upgradation

Physical survey
Organisation of community and community leaders
Marking of housing unit numbers
House count survey
Socio-economic survey
Preparation of draft list of dwellers
Finalization of list of dwellers after display and objections
Preparation of attribute database
Preparation of spatial database
Preparation of amelioration plans
Preparation of PC-1 of the development plan and its approval
Preparation of application forms and stationery
Issue of notice for inviting applications for regularization of area under housing units
Distribution of application forms for regularization of area under housing units
Receipt of applications along with payment
Scrutiny of applications
Issue of offer for regularization of area under housing units
Recovery of cost of land under housing unit
Implementation of development plan
Transfer of legal rights.

Rehabilitation (MUSP-Farash)

Organisation of community and community leaders
Marking of housing unit numbers
House count survey
Socio-economic survey
Preparation of draft list of dwellers

Finalization of list of dwellers after display and inviting objections
Scrutiny of final list by AMT, NAB, and Directorate of S&I, CDA
Verification of pending cases
Preparation of attribute database
Preparation of layout plan of proposed site of relocation of katchi abadi
Approval and demarcation of approved plan
Design of infrastructure facilities
Preparation of cost estimates and its approval
Preparation of PC-1 of the development plan and its approval
Reference to the government for relaxation in the provision of Islamabad
    Land Disposal Regulation, 1993
Preparation of application forms and other documents
Issue of notice for inviting application for allotment of plots
Distribution of application forms to clear housing units
Award of development works
Levelling of roads and streets (earthwork)
Metalling of roads and streets to the extent of 10 feet
Laying of drainage/sanitary system
Installation of water supply through community hand pumps
Development of site office
Payment of tentage area
Construction/development of temporary public toilets and septic tank
Receipt of applications along with down payment
Scrutiny of application forms
Balloting for allocation of plots
Issue of offer of allotment of plots to the eligible applicants
Shifting of dwellers to rehabilitation site
Start of construction of houses
Inauguration of project by the chief executive of Pakistan
Recovery of total cost of plot
Transfer of legal rights.

Source: Husain (2002: 20).

When CDA started survey and mapping of the KAs, it could not proceed any further because the federal government had put a ban on hiring staff. CDA is an extremely well-financed organization but it could not spare the meagre resources needed for survey and mapping. Such idiosyncrasies are part of working with the government. Tasneem Siddiqi used his good offices and put CDA in touch with UNDP's small grants programme Local Initiative Facility for Improvement of Urban Environment (LIFE) to seek a grant for this purpose.[10] With LIFE's grant, a preliminary survey of all KAs was undertaken. The key issues included taking count of existing KAs and determining the KAs that could be regularized and the ones that needed to be relocated; deciding a cut-off date after which no claim for a house built could be entertained; surveying the KAs, taking count of houses and inhabitants in each house and finalizing the list of potential applicants; mapping the land use pattern; and digitizing the collected information in the form of a GIS.

The survey revealed that 11 out of 31 informal settlements could qualify for regularization and upgradation in the first phase and the rest for relocation in the next phases. KAs not approved for regularization were recommended for relocation because they were currently situated at hazardous sites or on land allocated for amenity plots or prized land. Out of 11 KAs, six were notified for upgradation and improvement, and the rest were to be moved to a different location. Tasneem Siddiqi had also recommended that a new sector 1–14 should be developed for low-income residents to accommodate households evicted for relocation from unsuitable sites. There were other issues not anticipated by CDA as yet. Only the houses built up to 1995 were to be approved for regularization. Before the survey, Muslim Colony – the largest informal settlement of Islamabad with 30,000 population – which was first cleared off in 1979 for construction of Pakistan Institute of Medical Sciences (PIMS); and later 10 more KAs were officially recognized and added to the list. First survey was conducted with the help of an NGO with a cut-off date of 1995. Later one more KA was added and the cut-off date was extended to 1998 (Husain: 2002). An image of a KA (Figure 3.2) shows the contrast between the formal and informal housing in Islamabad. This required another survey by CDA. This time KAC took the lead in conducting the survey and verifying survey results in community meetings.

Figure 3.2 Squatter settlement in Islamabad
Source: Usman Qazi, *Express Tribune*, November 5, 2015

The survey consisted of a series of steps for data collection and verification. The first issue was determining the occupants settled before the cut-off date. It was a tricky issue. CDA wanted to ascertain how many families had settled at a location by a specific date to allot them a number and later on issue lease papers. However, as soon as a cut-off date was announced and the survey started, it created the hope that all the houses counted in the survey would qualify for signing the lease agreement. This tempted new families to move in and claim that they have been living there before the cut-off date. CDA had to develop an elaborate verification criterion to ensure that no intruder could make to the list. A procedure of reviewing evidence of support was strictly followed to verify the claims. The survey process started with a door-to-door survey, counting of families and family members in each home, preparation of a list, allocation of a house number and posting of the list at a public location. A few days after posting the list a community meeting was held to verify the list of families in the public meeting. Based on the discussion, the list was finalized and an alphanumeric number was allocated to each verified household. This process created many controversies because late entrants started using pressure through their patrons to be included in the list. In one case, when the director of KAC refused to enter a name in violation of the rules on the recommendation a highly influential official, he was suspended from his job. The director stood his ground and his list was finally accepted.

Once the list was finalized the map, household data, house numbers and other relevant information was digitized and converted into GIS. Six katchi abadis approved for regularization were located in sectors F-6/2, F-7/4, G-7/1, G-7/2, G-7/3 and G-8/1. These KAs were to be upgraded in an organic way on an incremental basis. The remaining five katchi abadis in sectors I-9 (Essa Nagri), I-10/4 (Dhok Najju), I-11/1 (Miskeenabad and Afghan Basti), I-11/4 (Haq Bahu/Benazir Colony) and Muslim Colony were to be demolished and relocated. But in accordance with the National Policy on Katchi Abadis, 2001, "no dweller of *katchi abadis* is to be evicted without providing him alternate land." Communities and representatives of KAs were closely involved in the whole documentation and decision-making process. CDA had formed a steering committee for regularly taking stock of the situation and deliberating on the future course of action. It included the chairman of the CDA, the director of the KAC, the director general of SKAA, the coordinator of LIFE/UNDP and representatives of each KA as members.

During the same period the National Shelter Policy was approved by the president of Pakistan in 2001. The policy guaranteed right to shelter to all Pakistanis. Due to these developments, number holders in KAs were given a temporary paper in a series of public meetings with the promise that lease documents would be issued to each number holder in due course. In the meanwhile, CDA sent all supporting documents along with the recommendations that these residents should be issued lease documents to the Ministry of the Interior. It seems that the scene was now set to start regularization and relocation of KAs with the community engagement. During the ensuing period, due to the change of government, the separation of powers between federal and provincial government and the restructuring of federal ministries, final approval on conferring lease title to the qualified dwellers of KAs was put on hold. Chairman Qamruzzaman Chaudhry was also transferred. Due to these changes final approval for regularization was never received from the parent Ministry of CDA.

From survey to settlement

During the process of documentation and digitization of data on katchi abadis it was decided to pick two KAs, one for upgradation at existing location and the other for relocation and rehabilitation; the katchi abadi in G-8/1 was notified for upgradation on a self-finance basis. Upgradation was meant to keep the organic structure of the settlement without any major alterations. At some places residents were asked to voluntarily move their house boundaries to create more space for the lane but the settlement was not restructured on a grid pattern. This was done to avoid any sharp increase in the price of plots and commercialization of KA land. KA G-8/1 was divided into of 575 plots measuring 20 × 30 feet. The plots were offered for sale to the residents at a subsidized rate of Rs. 40,000. In the first phase, 105 plots were allotted at the down payment of Rs. 15,000 in 1996 and possession was handed over on receipt of down payment. The remaining Rs. 25,000 were to be recovered in monthly instalments of Rs. 500 over a period of four years. The rate of default in monthly instalment payments was 80 percent. The second phase was interrupted due to a Lahore High Court Stay Order obtained by the residents of neighbouring sectors against regularization of the settlement. CDA succeeded in getting the stay order vacated, a revised development plan was prepared in line with the requirements of the neighbouring residents and the upgradation work was resumed.

Muslim Colony was selected for relocation and rehabilitation. CDA decided to finalize a Model Urban Shelter Project (MUSP) for the relocation of residents of Muslim Colony from Nur Poor Shahan to Ali Pur Farash by the end of April 2001. This project for upgradation and rehabilitation was initiated in September 1999 and implemented in six phases with a budget of Rs. 2.709 million for technical assistance from UNDP.

The CDA-SKAA collaboration turned out to be a successful and truly unique partnership, as CDA acquired all the required technical assistance from another government agency, SKAA, and achieved a landmark change in CDA policy and practice without any foreign technical experts (Husain: 2002). This success was considered very significant by UNDP and a short case study of the project compiled by UNDP's Deputy Resident Representative (DRR) Lena Lindberg was submitted to UNDP's knowledge management hub in Oslo.

Initial data collection work in the colony started in January 2001 and selection and development work of the new site, Farash, was also taken up. The work of rehabilitation of the dwellers of Muslim Colony started on an urgent basis in February 2001 along with the work of katchi abadi F-6/2. Minister of Environment, Local Government and Rural Development Omar Asghar Khan took personal interest in mainstreaming the lessons learnt from CDA experience into the National Shelter Policy and engaged key players as a member of the Local Government Advisory Board. The director general of SKAA and the national coordinator of LIFE were made members of the board. The case for awarding ownership deeds to the rest of the KAs was moved for approval to Minister of Interior Afteb Sherpao. The work on subsequent upgradation was later interrupted and CDA started eviction of some KAs in violation of the National Shelter Policy and the Islamabad High Court decision.

Conclusion

A review of long history of shelter problems in Islamabad clearly shows that the problem can be solved within the given means and the existing system. Though Doxiadis's plan did not take into consideration the needs of the poor while preparing the Master Plan for Islamabad, lack of a solution to this issue cannot be blamed on his faulty planning. CDA during the course of its work gained valuable insights about solving this problem, but due to institutional limitations no amicable solution has been implemented. As Akhter Hameed Khan used to say, "you don't replicate programmes, you replicate people". Islamabad needs to wait for another pioneer to pick the thread from where it was dropped in the beginning of the millennium.

Notes

1 Construction of unauthorized mosques is partly a poor man's way of solving the job creation and housing problem.
2 Officially there are no street dwellers in Islamabad, but street dwelling is taking place in imperceptible way. I have seen daily wage earners sleeping in the corridors of sector markets in civic centre G-6/2 and F-7/2. They keep their beds in the nearby mosques or the corner of a makeshift tea stall, spend the day time either on the streets or the sector's unkempt park in the shopping area, occasionally take baths in the mosque and use toilet facilities in the mosques; most of them regularly offer prayers. In F-7/2 a local philanthropist regularly provides a free meal to these workers in the evening. Free meals are also offered at the shrines of Pir Mehr Ali Shah and Bari Imam and by many private philanthropists at various points in Islamabad. I approached a local newspaper to write about the plight of these homeless workers and was informed that if a story on their miseries is published, CDA staff will start action against them and add to their misery. So, the principle is 'see no evil and say no evil'.
3 See the report by All-Pakistani Alliance for Katchi Abadis (2016).
4 An Orwellian explanation seems to be the best one on the issue of shelter in Islamabad.
5 Julius Salik, Minister of Minorities in the People's Party's government, and Omar Asghar Khan, Federal Minister of Local Government, Rural Development and Environment under the Musharraf government were vocal supporters of the residents of katchi abadis in Islamabad. While Julius Salik resorted to dramatic protests and seeking promises from the government for allocation of plots for the poor, Omar succeeded in getting the National Shelter Policy approved to provide a basis for a systemic solution.
6 See Federal Ombudsman Draft Report on CDA, mentioned in the Report on Proceedings of the Citizens Forum on May 16, 2013. The report gives us a detailed account of rampant corruption and mismanagement relating to land development and neglect of the poor inhabiting in informal settlements. www.mohtasib.gov.pk/wafaqimoh/userfiles1/file/Interim%20Report.pdf.

7 It is pertinent to note here that four UN agencies including UNDP and UNFPA moved their offices to the F-7/2 residential area in a multistory, purpose-built facility built at their request in complete violation of zoning laws. UNDP's DRR Operations came up with the brilliant idea of using the green belt as the parking lot in complete violation of environmental concerns. Residents of the lane, including one former UNICEF staff member, went to court to pre-empt the use of the building. This move had created a traffic and safety hazard for the residents in addition to disturbing the peace and calm of the neighbourhood. The contractor defended the case on the ground that the building under construction was meant for widows and orphans.

8 Federal Ombudsman Draft Report on CDA, mentioned in the Report on Proceedings of the Citizens Forum on May 16, 2013.

9 UNDP: Report on Proceedings of LIFE Partners Workshop, Faisalabad, Pakistan 1998 (in Urdu).

10 LIFE was established in pursuance of Local Agenda 21 at the Earth Summit in Rio in 1992. LIFE was mandated to provide grant funding to community-based organizations (CBOs), NGOs and local government authorities for improvement of water, sanitation and solid waste management and other activities relating to improvement of environment. The programme started operation in Pakistan in August 1993.

References

Adeel, M., 2010, "Role of Land Use Policy behind Unauthorized Expansion in Islamabad", *46th ISOCARP Congress 2010*, Nairobi, Kenya.

Ahmad, Niaz and Ghulam Abbas Anjum, 2012, "Legal and Institutional Perplexities Hampering the Implementation of Urban Development Plans in Pakistan", *Cities*, Vol. 29, pp. 271–277.

All-Pakistani Alliance for Katchi Abadis (APKAA), 2016, *Islamabad Katchi Abadi Community Database*, viewed from www.participedia.net/en/cases/islamabad-katchi-abadi-community-database.

Baig, M.A., 2017, "Pakistan's Real Estate Divide – Trends – Aurora", *Dawn*, viewed on January 7, 2017, from http://aurora.dawn.com/news/1141727.

Beacco, D., 2018, "Urban Planning in Islamabad: From the Modern Movement to the Contemporary Urban Development Between Formal and Informal Settlements", in A. Petrillo, and P. Bellaviti (eds.), *Sustainable Urban Development and Globalization: Research for Development*. Springer, Cham.

Daechsel, M., 2013, "Misplaced Ekistics: Islamabad and the Politics of Urban Development in Pakistan", *South Asian History and Culture*, Vol. 4, No. 1, pp. 87–106. Online viewed from http://dx.doi.org/10.1080/19472498.2012.750458.

Doxiadis, C. A., 1965, "Islamabad, The Creation of a New Capital", *The Town Planning Review*; Vol. 36, No. 1, ProQuest pg. 1.

Hasan, A., 2002, "The Changing Nature of the Informal Sector in Karachi as a Result of Global Restructuring and Liberalization", *Environment & Urbanization*, Vol. 14, No. 1.

Hull, M., 2009, *Uncivil Politics and the Appropriation of Planning in Islamabad in Naveeda Khan Crisis and Beyond: Re-evaluating Pakistan*, Routledge, London.

Husain, T., 2002, *Katchi Abadis and Some Viable Alternatives: A Case Study and Operational Guidelines Based on The Capital Development Authority, Islamabad's Approach, 1998 to 2000 UNDP Local Initiative Facility for Urban Environment (LIFE)*, Rural Support Programme Network (RSPN).

Kreutzmann, H., 2013, "Planning and Living in Islamabad: Master Plan Changes and People's Actions Are Transforming Pakistan's Capital", in Khalid W. Bajwa (ed.), *Urban Pakistan: Frames for Imagining and Reading Urbanism* (pp. 127–142), Oxford University Press, Karachi.

Mahsud, Ahmed Zaib Khan, 2008, "Doxiadis' Legacy of urban design: Adjusting and amending the Modern", *Ekistics*, Vol. 73, No. 436/441, pp. 241–263.

Mahsud, Ahmed Zaib Khan, 2007, "Representing the State. Symbolism and Ideology in Doxiadis' Plan for Islamabad", in Mark Swenarton, Igea Troiani and Helen Webster (eds.), *The Politics of Making* (pp. 61–74), Routledge, New York.

Malik, Sana Shahid, 2017, *The 'unplanned' Islamabad: State and Evictions in I-11 Sector*, master's Thesis submitted to Central European University, Department of Sociology and Social Anthropology, Budapest, Hungary.

Moatasim, F., 2015, *Making Exceptions: Politics of Nonconforming Spaces in the Planned Modern City of Islamabad*. PhD dissertation.

Moatasim, F., 2017, *Encroachment by the Elite in Bani Gala: How Islamabad's Posh Bani Gala Neighbourhood Is Built on Illegalities*, viewed from www.area148.com/encroachment-elite-bani-gala/.

Sayeed, Asad, Khurram Husain and Syed Salim Raza, 2016, *Informality in Karachi's Land, Manufacturing, and Transport Sectors: Implications for Stability*. Washington, DC: United States Institute of Peace.

Zaidi, S. Akbar, 2001, *Can the Public Sector Deliver? An Examination of the Work of the Sindh Katchi Abadis Authority*. City Press for UNDP/LIFE Pakistan, Karachi.

4 Women cotton pickers in Pakistan

Lost between the civil society and the state

Background and theoretical context

State, civil society and communities

Human development in developing economies is seen as a problem of accessing opportunities and realization of human freedom (Sen: 1999). Civil society intervention is presented as the ideal path to find access in case of market and public policy failure. However, civil society also has its share of failure in providing access for human development to vulnerable sections of society, especially women. This calls for a critical appraisal of the prevalent notion of civil society, state and communities. Community choices in post-colonial societies are made through a matrix of relationships defined by the modern nation state, markets, civil society and communities. There are divergent theoretical views on the nature of state and civil society in post-colonial societies. Despite sharp disagreements, there is a convergence of opinion that the state is inaccessible to communities (Patel et al.: 2001). A classical Marxian framework builds the case for inaccessibility of the state by presenting it as an instrument of exploitation and oppression of one class by another (Marx and Engels: 1969). Post-colonial theorists make the case for inaccessibility of the state due to its overdeveloped structures in relation to underdeveloped societies (Alavi: 1972, 1982). Human development practitioners point out that the post-colonial state is not accessible because it has inherited a developed law and order administration but lacks a development administration (Khan: 1980, 1996). Leading development assistance agencies are of the view that the state is inaccessible due its weak enforcement capacity, underdeveloped technical capacity or limited capability for realizing the 'trickle-down' effect (UNDP: 2015, Myrdal: 1968). Economic experts argued that post-colonial states deny access to low-income communities because they lack resources to realize their development objectives (Haq: 1966).

This chapter argues that civil society organizations can help low-income communities, especially women, access public sector resources only if they can engage both the state and civil society (NCSW: 2012). Rights-based talk, legislation and budgeting cannot produce results unless civil society organizations (CSOs) create support systems that help women to claim rights owed them by the state and the business sector. Deprivation of working class women, specifically cotton pickers, is not due to lack of resources to protect them from hazardous work conditions but to the absence of a support system to provide them this protection (Easterly: 2001). In the case of women cotton pickers, international conventions, state authorities and CSOs have not been able to develop instruments to help cotton pickers claim their rights. This argument is supported by the evidence of a large philanthropic sector, double the size of public sector development funds (AKDN: 2000), and the existence of a large informal sector with a significant asset base (Hasan: 2016) in tandem with the denial of means for protecting cotton pickers from fatal health hazards that are part of their work environment.

The state supports communities through provision of legal, fiscal and administrative space for their development. The resources allocated and legal and fiscal space conceded by the state to the low-income communities have to a large extent not been appropriated by the civil society for full realization of human development potential in Pakistan (Mustafa: 2005; Sattar: 2011; Hasan: 2016; Cheema: 2006). The state's expressed intent does not result in desired outcomes in the case of women cotton pickers due to unregulated agriculture sector and lack of a development administration to address the human development issues. The vacuum created by absence of a development administration was traditionally filled by local chiefs who used feudal authority to command unpaid labour for construction and maintenance of public works, settle community disputes and attend to matters of community concern. With the passage of time, feudal economic and social authority collapsed. Feudal authority was replaced either by mafias or by the civil society. Mafias built connections with lower tiers of government and encroached upon public resources to provide access to communities in return for payment of rent. CSOs succeeded in creating access for low-income communities and women only by developing knowledge tools that would enable them to negotiate on favourable terms with the government (Khan and Khan: 2004). This chapter assesses why efforts by various CSOs have not had much success in providing protection to women cotton pickers.

Agriculture in Pakistan: in need of going back to the future?

Pakistan inherited a pristine and pure agricultural production system. All major agricultural crops including wheat, cotton, rice, vegetables and fruit were grown as organic crops. There was no use of chemical fertilizers or insecticides up until the advent of the Green Revolution in the 1960s. The introduction of high-yielding varieties of seed and intensive water and fertilizer use was unfortunately accompanied by indiscriminate use of poisonous chemicals by local farmers, and these substances entered the food chain and agricultural production system. The term 'Green Revolution' was first used by USAID Administrator William Gaud in a speech on March 8, 1968. The Rockefeller Foundation and the Ford Foundation played important roles in developing this package, and USAID helped introduce it all over the world. High-yield varieties of wheat and cereals and later on genetically modified varieties of cotton were introduced to alleviate poverty on a mass scale. This role was acknowledged when Norman Borlaug, 'the Father of the Green Revolution', was awarded the Nobel Prize in 1970. Ironically these technologies had very adverse financial and health consequences for the poor. An artistic depiction of the impact of green revolution is given in a painting shown in Figure 4.1. Initially, the Green Revolution sharply increased wheat production, which plummeted due to land degradation within a decade. The government which introduced this technology was overthrown due to mass protests against the high price of wheat in the late 1960s.

Impact of these new practices on land, the health of farmers and male and female field workers engaged in production cycle was not given much consideration. Most of these farmers were illiterate. They could not even read the labels on fertilizer bags. They were not given any instructions on the use of these chemicals and precautions they needed to take to protect themselves in dispensing them.

Figure 4.1 Green Revolution

Source: Painting by Ijaz ul Hassan Mian

Due to deregulation of pesticides and insecticides in the 1990s, the whole cotton crop cycle was contaminated (Hussain: 1999). The use of hazardous chemicals for increasing 'productivity' in agriculture played havoc with the lives of women in agriculture. Contrary to the common perception that women are confined to the four walls of their house in a Muslim society, women play an active and equal part with men during the whole production cycle of agricultural commodities. Due to the prevalence of a barter economy, women labour was counted as 'unpaid', but under the customs and norms of the sharecropping agriculture they had to be compensated in some way, even if on highly unequal terms. While the international development community, UN agencies and IFIs were focusing their attention on collection of gender disaggregated data, counting women labour as part of GDP calculation, gender mainstreaming at workplace, safety and security at work, advocacy for greater budget allocation for health and education, the plight of women cotton pickers received little attention other than in the production of reports and papers.

Pakistan has ratified or acceded to all major human rights treaties including the Convention on the Rights of the Child (CRC), the Convention on Elimination of All Forms of Discrimination Against Women (CEDAW), the International Convention on Economic, Social and Cultural Rights (ICESCR) and the International Covenant on Civil and Political Rights (the ICCPR) in the spirit of inclusive governance. Pakistan is a signatory to almost every major treaty or convention relating to human and social development. Pakistan also committed to achieving both the Millennium Development Goals (MDGs) and Sustainable Development Goals (SDGs). Experience in Pakistan has shown that neither generous allocation of resources nor legislation and raising voices can lead to improvement of livelihoods unless three other conditions are met: (1) formulation of development plans based on sound documentation of local knowledge; (2) engagement of communities of interest with proper tiers of government; and (3) mobilizing a strategic coalition of partners to influence decision makers at the highest level. Due to absence of these conditions, ratification of global treaties has been accompanied by deregulation of pesticide import and sale in complete disregard of the farm workers.

To deal with the policy failure arising out of pesticide deregulation, European Union and international NGOs used a market-based approach to discourage the indiscriminate use of pesticide. This approach did not succeed either. These policy and market failures have been accompanied by civil society failure, because no CSO in Pakistan has developed tools which may enable communities to negotiate with cotton farmers to seek a hazard-free work environment. During the last four decades, only one Pakistani NGO, the Sindh Community Foundation, started unionizing women cotton pickers. They organized their first strike in 2014 which led to increased wages for 40 union members. This event was of course proudly publicized by the International Labour Organization, the media and NGOs. Despite this landmark achievement, no significant work has been done to protect women cotton pickers and their children against the perils of intensive, indiscriminate and irrational chemical use in the cotton crop.

State, civil society and communities

It is pertinent to note in this respect that state-civil society relations constitute three distinct subspaces: cooperation, collaboration and confrontation. When state and CSOs have common goals and means they cooperate; when they have similar goals and dissimilar means they collaborate; and when they have dissimilar goals and means they confront (Leitner: 1882; Najam: 2000; Baqir: 2007). Support to women cotton pickers calls for civil society engagement with the communities of interest in the third subspace, because government's deregulation of insecticide and pesticide business (goals) and scanty extension services (means) is in conflict with the interest of common pickers to have regulated insecticide and pesticide business (goal) and awareness and enforcement of health hazards (means).[1] Both the ratification of international conventions and NGO advocacy have not helped in protecting women cotton pickers from the lethal impact of chemical poison that has found its way in the production and harvesting of cotton in Pakistan. This chapter reviews the relationship between the state, civil society and the communities in relation to realization of right of female cotton pickers to work in a safe environment in Pakistan in view of the evidence from the field.

The basic characteristic which distinguishes a CSO maintaining the status quo from a CSO that harnesses and unleashes the potential of local communities to realize a higher level of self-development is the means employed to achieve their goals. These means fall into two broad categories: (1) following a template produced in a context not related to the reality of poor or (2) initiating a process to help communities identify their priorities and develop their capacity to achieve them with their own efforts. The so-called right-based rhetoric and poverty alleviation jargon can be very deceptive because it emphasizes the goal but ignores the means. Raising voice to achieve these goals cannot bring any results if the right-based work is led by contractors and not facilitators. The course taken by a public, private or voluntary institution can be evaluated in terms of their impact on maintaining or changing the status quo, not by the claims they make.

The extent to which CSOs succeed in seeking access to state and market resources for human development is not hampered due to a power or resource deficit of CSOs but due to their undeveloped capacity to overcome a trust deficit with the key partners. In many cases we see the fiscal and legal space conceded by the state to the communities is not appropriated by the civil society and communities. The question is, what is the nature of this disconnect that characterizes the unclaimed space between the state and the civil society? Is it lack of heroism or lack of touch with ground reality? My contention is that in claiming the right to a safe work environment for female cotton pickers, "means do exist for the population to free themselves, but the option has remained undeveloped" (Sharp: 2011: 13), and that is the weakness of civil society and donor-funded NGOs engaged in poverty alleviation and right-based development.

Formal sector experts see the failure of public policy and the market as a failure of communities. Living conditions and strategies of survival of these communities are not seen as an expression of diversity but as lack of awareness, assets, finances and power of the people. On the basis of this vision, they consider it 'their burden' to provide handouts and subsidies to communities and guide them to seek their rights. They don't consider it important to gain an understanding of the people living in poverty and suffering from discrimination. They look at the people's lack of knowledge as illiteracy or ignorance.

They don't even know that low-income communities have successfully established their own water, sanitation, shelter, health, education and job creation systems. They meet their daily needs with their own resources and manage to survive under very adverse conditions. The limited capacity of the formal sector to help the poor can be gauged by looking at the size of the unspent social sector budget of the government, the number of ghost facilities in education and the health sector and the magnitude of unutilized or misappropriated funds in the public sector. The formal sector does not even have maps of most of the settlements, and no documentation of informal economy exists at all.

What explains the difference between the success or failure of any intervention for uplifting the people is not the origin of these efforts in the public sector, market or civil society, but whether or not it enhances the range of choices of the people; whether this effort enables people to make the best use of their existing resources or limits their options to improve their quality of life. In the following sections we shall look at the issue of safe work environment of the female cotton pickers with reference to the key players in the economy.

Cotton economy of Pakistan

The cotton pickers

There are 500,000 cotton pickers in eight districts of Pakistan's cotton belt. Cotton accounts for 9 percent of value added in agriculture, 2 percent of GDP and indirectly contributes to one tenth of GDP and two thirds of merchandise export from Pakistan. Cotton is grown on more than 3 million hectares and female agricultural workers pick 1 million tonnes of cotton every year (Siegmann and Shaheen: 2008: 619–620). A majority of the farmers cultivate large plots varying between 10 and 400 acres, employing between 26 and 100 plus workers. It represents the largest share of employment in cotton production. Cotton pickers are mostly female seasonal workers, employed during the harvesting season, working in an unregulated sector. Women cotton pickers are not unionized and work for low or no pay in eight districts dominated by big landlords.

Market deregulation and pesticide use

Pesticide use was almost non-existent in Pakistan at the time of its creation in 1947. Pesticides were used for the first time in 1950s to combat locust attacks. In the 1970s, due to high pressure for increased production, pesticide use entered the production cycle and consumption of pesticides increased from 665 metric tonnes (MT) in 1980 to 69,897 MT in 2002. This led to a 90 percent decline in natural pest and insect enemies (Khan et al.: 2002). During the 1980s, heavy use of pesticides started in cotton fields in Punjab in response to the bollworm. As a result, the cotton crop became more pesticide intensive and cotton accounted for 80 percent of pesticide use until recently (Rana: 2010: 44). Part of the reason for wider use of pesticides is that local companies offer very lucrative incentives to pesticide dealers including cash prizes and trips to foreign countries. Consequently, overuse and dumping of discarded pesticide was also reported (Khooharo et al.: 2008: 69–70).

Poor work practices, a low level of toxic awareness, inadequate precautionary measures and inappropriate storage conditions have been noticed in cotton cultivation. Ineffective registration, monitoring and judicial oversight systems and ineffective control on sale of banned pesticides have further aggravated this situation (Feenstra et al.: 2000). Indiscriminate use of pesticide has had unintended consequences as well. Feenstra has reported the use of pesticide for committing suicide. The suicide pattern revealed that most of the victims were between the ages of 11 and 30, and 53 percent were females, mostly married women (Feenstra et al.: 2000: 10). A study conducted in 1990s on the knowledge of preventive measures for protecting against the side effects of pesticide showed that 80 percent of males and only 5 percent of females were aware of such measures.

Impact of pesticide on female cotton pickers

Cotton accounts for more than 54 percent of the pesticide used in Pakistan (Feenstra et al.: 2000; Khan et al.: 2002). More than 90 percent of cotton picking is done by women. Chemical pesticides are used indiscriminately, and pesticide residues enter the food chain. Use of pesticide at more than 10 sprays per crop is among the most alarming level. Traces of pesticide have been found in wheat, maize, rice, vegetable, pulses, cereals, meat, dairy products, poultry and animal feed. The same was the case with gram, and fruits. A study by Zia et al. has pointed out that some pesticides can persist in the environment for more than 20 years. A considerable number of samples showed more than the daily acceptable rate of chemical contamination. A high concentration of pesticide residues was found in cottonseed oil samples from different locations. Studies on accumulation of pesticide residues in human tissues and blood found between 70 percent and 100 percent people were affected (Feenstra et al.: 2000; Khan et al.: 2002; Tariq et al.: 2007; Zia et al.: 2009; Khan et al.: 2011). Blood samples of cotton pickers in Multan found 82 percent to 88 percent of respondents containing pesticide residues. "Analysis of blood sample of women cotton pickers showed high level of headache and muscular weakness as a common trait among female cotton pickers. Researchers also found out that majority of farmers did not use any protective equipment" (Khan et al.: 2011: 185). After picking cotton, women reported complaints of sneezing, muscular pain, nausea, skin burning, itching, cough, headaches and blisters on their bodies (Feenstra et al.: 2000).

Most of these pesticides are suspected of being carcinogenic, mutagenic and teratogenic, linked with neurological disturbance, skin cancer, sterility, hypertension, hormonal disorders and blood disintegration. An estimated 700,000 cotton pickers, most of them women and girls, are employed on the 1.6 million cotton-growing farms in Pakistan during the picking season between September and December. The working environment of cotton pickers is full of poisonous pesticides. During the 8–9 hours of daily picking (Salman: 2007: 861; Khan et al.: 2007) acute and chronic poisoning takes place, leading to loss of work days and expenditure on treatment (Feenstra et al.: 2000; Khan et al.: 2007: 1127). Pesticide use suppresses the natural pest control processes as well (Khan et al.: 2007: 1132; Tariq et al.: 2007).

The results of blood analysis to detect level of poisoning among cotton pickers in post-pesticide spray period in Multan and Bahawalpur division showed that only 10 percent of females were found in the normal range of 88–100 percent whereas 42 percent were found to be in the hazardous range (Tahir et al.: 2001), Khan et al. found close association between pesticide exposure and disturbance of thyroid and reproductive hormone levels (Khan et al.: 2011; Siegmann: 2006; Abbas et al.: 2015). A study conducted by Abbas et al. reveals that while 54.5 percent of women experienced tiredness, 9.9 percent experienced mental disturbances and another 8 percent experienced fatigue. These women were not aware of the link between these problems and pesticide use. Only 8 percent of the women surveyed were aware of health hazards and only 10 percent wore protective clothing (Abbas et al.: 2015). The average age of cotton pickers is 30–33 years (39 percent), but minors also contribute to family labour to increase output. Illiteracy is very high (75 percent) among them and they are mostly from landless families (85 percent) and work as unpaid and low-paid family labourers (Haq et al.: 2008; Abbas et al.: 2015). In financial terms, the social and environment cost of pesticide use was calculated to be Rs. 11,941 million per year. Human health was calculated to cost Rs. 265 million. Women cotton pickers were reported to be spending Rs. 105 million on treatment of disease caused by pesticide exposure (Khan et al.: 2002).

Decision-making structures, policy and process

The policy-making structure on cotton is highly centralized, leaving no space for small farmers, field workers and vulnerable female cotton pickers to have their voice heard at the policy-making forums. Referring to the government's policy-making, Rana argues that multi-national corporations (MNCs), government agencies, government seed producers, private seed producers, farmers and the textile industry are the key stakeholders in policy-making. The government prefers to monopolize policy legislation, accommodates the interests of MNCs due to the expressed interests of textile industry and shows indifference to the farmers' interest (Rana: 2010: 159–169). Handing over pesticide distribution to the private sector in 1980 and deregulation of pesticide registration in 1993 was meant to appease the MNCs. Decision-making structure is clearly biased in the interests of powerful commercial, industrial and big farmer lobbies. This has led to policy failure in protecting the safety of female cotton pickers. This policy failure combined with market failure has had detrimental effects on the health of female cotton pickers, their children and families.

Public policy on pesticide use

Public policy did not address the issue of enormous divergence between the private and social costs of pesticide use. In 1980 pesticide distribution was handed over to the private sector and in 1993 pesticide registration was deregulated (Khan et al.: 2002). It led to excessive and indiscriminate use of pesticide, transferring the negative cost of pesticide use to farm workers, cotton pickers and consumers of farm products and led to ecological and environmental degradation. Rana points out that policy-making is characterized by internal tensions between various power groups and the division of policy-making responsibility between various actors within the government. He emphasized the importance of democratic over autocratic or technocratic processes of policy-making (Rana: 2010: 29) and the technology-society nexus (Rana: 2010: 2).

Post-1980 legislation transferring import and sale of pesticide from the public to the private sector led during the next two decades to a 70-fold increase in pesticide consumption, 80 percent of which was used for cotton crop. Punjab used 90 percent of the pesticide (Khooharo et al.: 2008: 57–58). The import liberalization policy of the government led to a massive increase in the import of hazardous pesticide by the private sector. Under this policy, import of generic compounds registered outside Pakistan was allowed without field testing. Environmental and social cost of indiscriminate import and use of pesticide was estimated to be US$206 million per year. In addition, it was noted that use of chemical pesticide increased the pest problem, disturbed the eco-agricultural system and drained the national exchequer (Khan et al.: 2007: 1119). Despite the critical nature of these challenges, it has been noted that women are invisible in national agriculture policy (Habib: 1997). A comprehensive national monitoring system is still missing. There is no system of social security (Pegler et al.: 2011: 28); voice or representation for improved labour processes is not allowed politically (due to lack of legislation); there is lack of coherent policies, an analytical framework and a monitoring mechanism (Tariq et al.: 2007).

Extension services

Extension services, which could have played a very effective role in preventing hazardous practices in cotton cultivation and harvesting, have been extremely ineffective. Both the public (Khan: 1980) and the private sector since 1970 (Rana: 2010: 45) provide extension services in Pakistan. Rural Support Programmes (RSPs) and NGOs are also engaged in extension work on a very large scale. A report by the Ministry of Finance has noted that

> Most private sector extension . . . is limited to the use of pesticides and herbicides. Other issues, such as farming practices, planting time, efficient irrigation, seed quality, plant density, integrated pest management, etc. are not considered relevant enough to feature in its advice. In addition, the public sector has a bias for educated farmers and the private sector for the large ones

(ibid). Both approaches are exclusionary in a country where the literacy rate in the rural areas is just 44% and where 58% of farms are less than 5 acres.

(Ministry of Finance: 2008)

Local seed companies work in an informal way, and extension services are provided by male extension workers (Siegmann and Shaheen: 2008).

Cotton trade value chain

Market failure can be best understood by looking at cotton production, processing and sale value chain. Value chain analysis can help identify the segment where worker safety and environmental regulations are being violated. It enables consumer unions and commercial buyers to stop purchasing cotton from farmers not complying with the norms for worker safety and environmental conservation. This innovative approach, followed by government bodies, international NGOs and socially responsible businesses, has discouraged business practices based on abuse of workers, production under hazardous conditions, and unfairly low wage payments to workers. This initiative, although very well intentioned, has not produced desired results because the agriculture market is not regulated and agricultural workers are not unionized. The value chain approach creates opportunities like ratification of human rights treaties and conventions, but their realization depends on presence of a support organization which can work in a non-threatening way in a very threatening social environment. Norms and practices of IFIs and global aid agencies do not allow them to invest in institution building for promoting work safety. That is where the fault sets in. However, it is important to look at the cotton trade value chain and identify the missing link in safe work environment. The cotton trade value chain, starting with cultivation, continues through harvesting, ginning (separation of cotton fibre from the seed), spinning, weaving (into cloth) and knitting (into garments).

Public sector institutions

The government structure consists of a number of institutions that influence agricultural policy decisions. These institutions include the Pakistan Central Cotton Committee (PCCC), Cotton Research Institutes (CRIs), the Pakistan Agricultural Research Council (PARC), the Textile Commissioner, the Agricultural Prices Commission (APC), the Federal Seed Certification Department (FSCD), various agricultural universities, and the Export Promotion Bureau (EPB). The country has 4,000 agricultural scientists, 500 agricultural extension agents, and proportional numbers of officials in seed certification and supply, agricultural machinery provision, policy development and agricultural pricing, and agricultural credit, all in the public sector. It is characterized by inertia on socially responsible policy-making and motivated by output maximization (Banuri: 1998).

Large scale importers

Large-scale importers of Pakistani cotton products include Walmart, Kmart, Sears, Dayton Hudson, Woolworth, J.C. Penney, May, Melville and The Gap in the United States; Bo Weevil in the Netherlands; Otto BV in Germany; Hennes and Mauritz in Sweden; and Stockmann in Finland. Their purchasing decisions are highly influenced by a very unequally structured cotton production system. The non-formal, non-mill sector contributes 90 percent of the fabric production but is not organized and carries no weight in comparison to the large-scale industrial sector. The same is the case with small farmers and women cotton pickers. On the other hand, the Pakistan Agriculture Pesticide Association (PAPA) is a very powerful lobbying group. Socially responsible buying has emerged as an important factor influencing the value chain in cotton but there is no information available on the impact of consumer-driven apparel chain on export of Pakistani fabric. But the research findings in neighbouring India are not very encouraging. A study by Ramamurthy on cotton farmers in Andhra Pradesh has noticed an increase in the wages of women and children (accompanied by de-schooling and longer working days than adults) on cotton fields in the short run due to globalization of trade, but a long-term trend of decline in real wages of women, deterioration of quality of life and increased indebtedness, despite the purchase of cotton from the consumer-driven apparel chain and their decreased dependence on rural elites. Price fluctuation has also led to suicide of hundreds of farm labourers due to deepening debt to pesticide dealers; some who did not commit suicide sold their kidneys (Ramamurthy: 2004).

Textile mill owners

Spinners are very well organized under the All Pakistan Textile Mills Owners Association (APTMA) which represents 360 mills and is a lobby with strong political clout. They successfully resisted quality-based pricing of cotton which would have allowed a premium for contamination-free organic cotton (Pegler et al.: 2011). They have not invested a single penny under corporate social responsibility (CSR) to improve the working condition of cotton pickers. They are so powerful that they have received significant benefits from international trade without meeting their human rights obligations. In 2013 European Union decided to grant Generalized Scheme of Preference (GSP) Plus status to Pakistan. It allowed almost 20 percent of Pakistani exports to enter the EU market at zero tariff and 70 percent at preferential rates. It created an opportunity for the textile industry to earn profits of more than Rs. 1 trillion per year. Despite the EU's and the Pakistani government's commitment to numerous international human rights treaties and conventions no regulatory body was formed to screen textile exports originating in the enterprises engaged in exploitation of the workforce and neglecting worker's safety. The Government of Pakistan did not have any labour protection laws and monitoring mechanisms in place to protect workers' rights in textiles and agriculture under GSP (*Daily Times*: 2014).

Pesticide dealers

In 1980 pesticide distribution was handed over to the private sector, and in 1993 pesticide registration was also deregulated, giving a free hand to pesticide dealers to promote their product among the farmers without requiring them to take precautionary measures to guide them about the hazards of indiscriminate and irrational use of their products and introducing the guidelines to protect their health (Khan et al.: 2002).

Farmers

Pakistan's cotton sector is characterized by significant diversity in terms of size of land ownership, means of access to technical knowledge and connection with corporate importers. The response of different players to policy instruments and market incentives has varied with their position in the power structure. Yarn production is in the large-scale organized sector with a powerful lobbying capacity; the weaving sector is very anarchic without a coherent collective structure, and small-scale farmers (1.3 million) have no organization and possess no power over market prices. The textile and apparel industry is large scale and retailer driven – they manage trade and production networks and are concentrated in few hands. They have developed closer inspection and transaction arrangements with small and decentralized producers. The cotton-producing area in Pakistan is dominated by big landlords who are spiritual leaders, elected representatives, ministers and landholders at the same time. Poorly paid, disorganized and vulnerable cotton pickers are at the receiving end in dealing with their irrational and harmful agricultural practices. A study by Khan showed that while number of pesticide sprays increased from 0–1 per crop in 1980s to 7–11 per crop in late 1990s, pesticide use could be reduced considerably without reducing the cotton yield. Inefficient farmers were spending 70 percent more on pesticide compared to best farmers and producing the same level of output. However, no institutions exist that can teach these farmers what is in their own best interests (Khan et al.: 2002).

Dynamics of the market and the plight of cotton pickers

Due to lack of unionization, lack of a support system and existence of an unregulated agricultural sector, opportunities offered throughout the value chain by global market initiatives have not been availed by the cotton pickers in Pakistan.

Kings Apparel, a knitwear production company, initiated an organic cotton project in Pakistan in January 2000 (Kings Apparel: 2005). It was a market-based alternative to deal with public policy failure. The assumption of the intervention was that a subsidy reduction at the international level and the resulting price increases for raw cotton may support higher wages for the cotton pickers and provide them a safe work environment as well (Siegmann and Shaheen: 2008). This

opportunity could be realized under a regulated market by well-organized workers. Pakistan is a signatory to the ILO Conventions 100 and 111, which guarantee
the elimination of discrimination in respect of employment and occupation. But
under the national labour legislation, the agricultural sector is not even covered by
the Industrial Labor Ordinance 2001 (Ali: 2004). The market structure also does
not favour female cotton pickers. In this respect it is important to note that separation of ownership and labour, segmentation of work process and restriction of
various forms of labour to different social groups creates an oversupply of labour
in the field and creates downward pressure on wages in the global value chain
and diverts benefits gained from globalization to owners and contractors (Pegler
et al.: 2011: 5). Pegler has also pointed out that social profile, social regulation
and differentiation processes are important factors in determining entitlement for
employment and conditions of work. In the social differentiation process, gender,
age, geographic location and type of livelihood play a very important role. It is
instrumental in keeping the prices of cotton and textile products low in the international market.

There are 5,000 garment factories in Pakistan. Branded clothing produced in
Pakistan includes Nike, Kohl's, Sears, Walmart, The Gap, Old Navy, Macy's
and others (Stotz: 2015). The largest export destination for garments made in
Pakistan is the United States. The EU is also a main export destination for Pakistan: 72–75 percent of all Pakistani exports to the EU are clothes (Stotz: 2015:
21) "Pakistan has GSP+ status and thus it enjoys duty-free access to the European market" (Stotz: 2015: 8). About 20 percent of Pakistan's global exports
are covered by GSP+. Pakistan is a signatory to the International Covenant on
Civil and Political Rights (ICCPR), the International Covenant on Economic,
Social and Cultural Rights (ICESCR), the Convention on Elimination of All
Forms of Discrimination Against Women (CEDAW) and the Convention on the
Rights of the Child (CRC), but it has expressed reservations about various arti-
cles and follows its own interpretations in this regard. In addition, Pakistan has
ratified all of the following ILO conventions: C029 – Forced Labour Conven-
tion, C087 – Freedom of Association and Protection of the Right to Organise
Convention, C098 – Right to Organize and Collective Bargaining Convention,
C100 – Equal Remuneration Convention, C105 – Abolition of Forced Labour
Convention, C111 – Discrimination (Employment and Occupation) Convention, C182 – Worst Forms of Child Labour Convention and C1338 – Minimum
Age Convention. When ratifying C1338, Pakistan specified the minimum age
to be 14 years. Pakistan has ratified these treaties but not entered into any
agreement, that is they have not signed the operational protocols that make
individual complaints possible, and there is no way to enforce these rights
(Stotz: 2015). The plight of women cotton pickers also relates to ILO's goal
for Pakistan to implement the Decent Work Agenda. However, cotton pickers
have not been able to avail any protection from harm under the provisions of
this commitment. Barriers in seeking justice include little knowledge of legal
rights, lack of organization, fear for safety (i.e. protection against violence),
the high cost of legal aid, corruption and ineffective judgements (Stotz: 2015).

Rural women in the cotton belt have much restricted social, occupational and geographic mobility, and are confined mostly to low-paid and seasonal work. Picking is done in groups of 5–25 workers formed by contractors. Women are paid by the weight of their output. Cotton pickers are given a receipt for the cotton collected at the end of the day and paid at the end of the season. This payment is received by the male members. A fast picker can pick up to 1 *maund* (40 kg) of cotton a day. Estimates of payment received for 1 maund vary between Rs. 40 and Rs. 85. Pickers associate their weak bargaining power with gender, as lack of female occupational mobility leads to oversupply of female labour during the picking season. Weight of the cotton harvested is also determined by the contractors who take advantage of lack of literacy and organization of female cotton pickers. Caste divisions play a major role in preventing unionization of women. Due to these divisions and fluctuating working hours during various stages of cotton picking, daily earning of the cotton pickers has been reported to range between Rs. 10 and Rs. 60. In certain instances, women reported walking up to four hours to reach the cotton fields. Due to exposure to poisonous substances generated by the pesticides, cotton pickers suffer from health hazards. It severely affects infants, as cotton picking is also common during pregnancy and breastfeeding. Pesticide residuals contaminate water, soil and the food chain, and the use of cotton stalks as fuel leads to respiratory, eye, skin and stomach ailments. Cotton pickers are not provided any guidance or protective gear to protect them from the hazards of pesticide use. Financial losses arising due to illness of cotton pickers causing work loss and treatment expenses have been estimated to be Rs. 765 million (Siegmann and Shaheen: 2008).

Civil society support to cotton pickers

There are many NGOs working on cotton pickers' rights, women empowerment, gender equity and safety at workplace for women, but they have done little for improving the quality of life for women cotton pickers, perhaps because it is not high on the donor agenda. The work of NGOs working for women rights ranges from research to advocacy to capacity building to demanding rights. Shirkat Gah (SG) – a leading documentation, research and advocacy centre on women's issues – alone contains a collection of 110,027 books, articles, annual reports and audio-visual material (CDs and DVDs; http://shirkatgah.org/documentation-centre/). As part of its model, it aims at narrowing the gap between the state and citizens. It states women's lack of knowledge concerning legal entitlements and existing services coupled with the paucity of non-family linkages as a major obstacle in claiming their rights. In 2011, SG replaced its past practice of working with rights holders and duty bearers in parallel streams with district core groups (DCG) in 42 districts that bring together local CSOs, duty bearers and service providers at one forum. SG started directly linking rights holders and civil society activists with both state officials and selected service providers. According to SG, DCGs have significantly increased access to rights and the responsiveness of duty bearers. Although SG works in eight core districts of the cotton belt, their website and list of publications don't provide any documentation on the impact of their work on cotton pickers. The same is the case with other research, advocacy and right-based NGOs. The cases of other CSOs regarding cotton pickers are similar.

Agency and resistance

Women's lack of access to centres of decision-making (Salman: 2007: 857) and NGOs' lack of knowledge and experience on organizing principles of women living under tyranny and techniques for negotiating with the employers and engaging the state does not take the work of women's emancipation very far. The state of the cotton pickers' working environment calls for engagement with three key players responsible for poisoning the cotton production cycle: government, pesticide distributors and farmers. Government needs to resume its role in regulating the import, distribution and use of pesticide; pesticide distributors need to take responsibility to educate their dealers to guide farmers in the rational use of pesticide and provide guidelines for protection of farmers, cotton pickers and field workers during farm work; and farmers need to comply with the guidelines for protection of farm workers and cotton pickers and provide them protective gear during the course of work. Scientific studies, data generated through field surveys and anecdotal evidence provided material which could be used for holding dialogue with the government, pesticide dealers and farmers. Some preliminary work was done by a corporate business and UN agencies to find alternative, chemical-free practices of cotton cultivation. But protection of female cotton pickers from the devastating impact of poisoning did not take place. Cotton pickers' agency and resistance is in its infancy and they have been abandoned by the state, the private sector and civil society.

Conclusion

In the struggle for human development, female cotton pickers suffer from a paucity of social capital. This deficit defines the space between the 'professional' outsider and 'illiterate' insider. There is no dearth of civil society networks that should be concerned with redressing the cause of female cotton pickers. These networks include networks of health workers, teachers, newspaper correspondents, extension workers and rights activists. However, none of these networks has lent its weight to the cause of cotton pickers. As noted by Salman, in the struggle to conserve ecology all those without market power (especially in view of privatization of commons), organization and knowledge become the losers (Salman: 2007: 856; Rana: 2010). Invisibility and indifference of the market is a lethal form of domination. Negotiating better terms of employment requires leadership familiar with the perceptions of cotton pickers and awareness of openings for engagement with the big cotton farmers.

Better terms of employment cannot be negotiated by writing reports or holding rallies. As noted by Salman, engagement of civil society with cotton pickers is missing:

> Pakistan is unique. While there are efforts on part of organizations towards mass awareness directed at and for women e.g. to preserve their forests and their cotton fields, the efforts are mostly donor-driven and not coming out of a true felt passion of the women themselves coming out to protest.
>
> (Salman: 2007: 862)

An alternative tried by UN agencies was to start working with farmers through farmer field schools (FFS) for integrated pest management (IPM). These schools were successfully tested with the help of UNDP-FAO in Vehari district of Southern Punjab and supported for replication as the Vehari model by UN agencies (Khan et al.: 2007: 1119). Khan et al. noted that farmers' education without hands-on learning through FFS did not contribute to improved decision-making and pesticide use practices. They suggested more involvement of plant protection experts during FFS training for enhancing the understanding of farmers, extension workers and researchers (Khan et al.: 2007: 1127, 1134).

There are trailblazing women who have fought for causes of women in other areas; for example women living in Karachi Administrative Women Workers Society (KAWWS) approached the court for improvement of sanitation under the leadership of Safina Siddiqi and won the case. Judge allowed KAWWS members to stop paying tax to the municipal government until satisfactory services were provided. Perween Rahman secured legal titles for thousands of low-income families living in urban and peri-urban areas of Karachi through her rigorous documentation and mapping work. Shahla Zia was able to win a case against construction of an electricity grid station in a neighbourhood in Lahore on designated green belt property through public interest litigation. The court considered the case to be maintainable under Article 184(3) since the danger and encroachment alleged were such as to violate the constitutional right to life when interpreted expansively. Nothing like that has happened in the cotton fields (Tariq et al.: 2007). Numerous CSOs have raised voice for the cause of cotton pickers' rights (Bhatti: 2015). But walking the talk is more critical than raising the voice in five-star hotels (Cusack: 2015).

Note

1 Communities can be defined with reference to their common interest, location, beliefs or values. For the purpose of this chapter, we consider the relationship between civil society organizations (CSOs) and communities.

References

Abbas, Mazher, Irfan Mehmood, Arshed Bashir, Muhammad Ather Mehmood and Sonila Hassan, 2015, "Women Cotton Pickers' Perceptions about Health Hazards Due to Pesticide Use in Irrigated Punjab", *Pakistan Journal Of Agricultural Research*, Vol. 28, No. 1.

AKDN Pakistan, 2000, *Philanthropy in Pakistan*, AKDN Report.

Alavi, H., 1972, "The State in Post-Capitalist Societies: Pakistan and Bangladesh", *New Left Review*, Vol. 74, No. 1, pp. 59–81.

Alavi, H., 1982, "State and Class under Peripheral Capitalism", in H. Alavi and T. Shanin (eds.), *Introduction to the Sociology of 'Developing Societies'* (pp. 289–307), Macmillan Education Ltd., London.

Ali, K., 2004, "Labour Policy in Pakistan", in K. Ali (ed.), *Sustainable Development: Bridging the Research/Policy Gaps in Southern Contexts*, Oxford University Press, Karachi.

Banuri, Tariq Pakistan, 1998, *Environmental Impact of Cotton Production and Trade*, International Institute for Sustainable Development, Winnipeg.

Baqir, F., 2007, *UN Reforms and Civil Society Engagement*, UNRCO, Islamabad.

Bhatti, Rubina F., 2015, *Exploring Strategies for Effective Advocacy: The Lived Experience of Leaders of Pakistani Non-Governmental Organizations*, PhD dissertation, University of San Diego.

Cheema, Ali and Asad Sayeed, 2006, *Bureaucracy and Pro-poor Change*, PIDE Working Papers.

Cusack, John and Arundhati Roy, *That Can and Cannot Be Said: A Conversation between John Cusack and Arundhati Roy*, Outlook, Op-Ed, viewed on November 16, 2015.

Daily Times, 2014, *Government Should Provide Legal Cover to Labour of Women Cotton Pickers*.

Easterly, W., 2001, *The Political Economy of Growth Without Development: A Case Study of Pakistan*, Development Research Group, World Bank.

Feenstra, Sabiena, Abdul Jabbar, Rafiq Masih and Waqar A. Jahangir, 2000, *Health Hazards of Pesticide in Pakistan: Report No. 100*, Pakistan Agricultural Research Council, International Water Management Institute (Pakistan Programme).

Habib, N., 1997, "Invisible Farmers-Rural Roles in Pakistan", *Pesticides News*, No. 37, pp. 4–5.

Haq, Qamar-ul-Haq, Tanvir Ali, Munir Ahmad and Farhana Nosheen, 2008, "Role of Age, Education and Landholding on Cotton Producing Community: A Case Study of District Multan", *Pakistan Journal Agricultural Science*, Vol. 45, No. 1.

Haq, Mahbub Ul, 1966, *The Strategy of Economic Planning: A Case Study of Pakistan*, Oxford University Press, Karachi.

Hasan, Arif et al., 2016, *Karachi: The Land Issue*, Oxford University Press, Karachi.

Hull, Matthew S., 2012, *Government of Paper: The Materiality of Bureaucracy in Urban Pakistan*, University of California Press, Berkley.

Hussain, Naila, 1999, *Poisoned Lives: The Effects of Cotton Pesticides*, Shirkat Gah, Women Resource Centre, Lahore.

Khan, Ayesha and Rabia Khan, 2004, *Drivers of Change Pakistan: Civil Society and Social Change in Pakistan*, Institute for Development Studies, Sussex.

Khan, Mohammad Ayub, 1967, *Friends Not Masters*, Oxford University Press, Karachi.

Khan, Muhammad Azeem, M. Iqbal and Iftikhar Ahmad, 2007, "Environment-Friendly Cotton Production through Implementing Integrated Pest Management Approach", *The Pakistan Development Review*, Vol. 46, No. 4, pp. 1119–1135.

Khan, Dilshad A., FCPS, Karam Ahad, Wafa M. Ansari and Hizbullah Khan, 2011, "Pesticide Exposure and Endocrine Dysfunction in the Cotton Crop Agricultural Workers of Southern Punjab, Pakistan", *Asia-Pacific Journal of Public Health*, Vol. 25, No. 2, pp. 181–191, APJPH.

Khan, Akhter Hameed, 1996, *Orangi Pilot Project, Reminiscences and Reflections*, Oxford University Press, Karachi.

Khan, M. Azeem, Muhammad Iqbal, Iftikhar Ahmad and Manzoor H. Soomro, 2002, "Economic Evaluation of Pesticide Use Externalities in the Cotton Zones of Punjab, Pakistan", *The Pakistan Development Review*, Vol. 41, No. 4, Part II, pp. 683–698.

Khan, Shoaib Sultan, 1980, *Rural Development in Pakistan*, Vikas Publicizing House Ltd., Ghaziabad.

Khooharo, Aijaz Ali, Rajab Ali Memon and Muhammad Umar Mallah, 2008, "An Empirical Analysis of Pesticide Marketing in Pakistan", *Pakistan Economic and Social Review*, Vol. 46, No. 1, pp. 57–74.

Kings Apparel, 2005, *Organic Cotton*, viewed from www.kingsapparel.com/organic.htm.

Leitner, G. W., 1882. *History of Indigenous Education in the Punjab since Annexation and in 1882*. Amar Prakashan, Reprint Delhi.

Marx, Karl and Frederick Engels, 1969, "Manifesto of the Communist Party", in Marx/ Engels (eds.), *Selected Works* (Vol. 1, pp. 98–137), Progress Publishers, Moscow.

Ministry of Finance, 2008, *Pakistan Economic Survey 2008–09*. Ministry of Finance, Government of Pakistan, Government Printing Press, Government of Pakistan.

Mustafa, Daanish, 2005, "(Anti)Social Capital in the Production of an (UN)Civil Society in Pakistan", *Geographical Review*, Vol. 95, No. 3, pp. 328–347, New Geographies of the Middle East.

Myrdal, Gunnar, 1968, *Asian Drama – An Inquiry into the Poverty of Nations* (Vol. 1), The Twentieth Century Fund, New York.

Najam, A., 2000, "The Four C's of Government Third Sector-Government Relations", *Nonprofit Management and Leadership*, Vol. 10, No. 4, pp. 375–396.

National Commission on Status of Women (NCSW) Annual Report, 2012, Opal Studio, Islamabad.

Patel, Sheela, Sundar Burra and Celine D'Cruz, (2001), "Slum/Shack Dwellers International (SDI)-Foundations to Treetops", *Environment & Urbanization*, Vol. 13, No. 2.

Pegler, L., K.A. Siegmann and S.R. Vellema, 2011, "Labour in Globalised Agricultural Value Chains in Value Chains", in A.H.J. Helmsing and S.R. Vellema (eds.), *Social Inclusion and Economic Development: Contrasting Theories and Realities*, Routledge, New York.

Ramamurthy, Priti, 2004, "Why Is Buying a 'Madras' Cotton Shirt a Political Act? A Feminist Commodity Chain Analysis", *Feminist Studies*, Vol. 30, No. 3, pp. 734–769, viewed on December 3, 2015, from www.jstor.org/stable/20458998

Rana, M.A., 2010, *Formalizing the Informal: The Commercialization of GM Cotton in Pakistan*. PhD thesis, Melbourne School of Land and Environment, The University of Melbourne.

Salman, Aneel, 2007, "Ecofeminist Movement-from the North to the South", *The Pakistan Development Review*, Vol. 46, No. 4, pp. 853–864.

Sattar, Nikhat, 2011, "Has Civil Society Failed in Pakistan?" *Social Policy and Development Centre (SPDC)*, Working Paper No.6.

Sen, Amartya, 1999, *Development as Freedom*, Oxford University Press, Oxford.

Sharp, Gene, 2011, *From Dictatorship to Democracy*, Serpent's Tail, London.

Shirkatgah, viewed on December 2, 2015, from http://shirkatgah.org/documentation-centre/.

Siegmann, Karin Astrid and Nazima Shaheen, 2008, "Weakest Link in the Textile Chain: Pakistani Cotton Pickers' Bitter Harvest", *The Indian Journal of Labour Economics*, Vol. 51, No. 4.

Siegmann, K., 2006, "Cotton Pickers after the Quota Expiry: Bitter Harvest", *SDPI Research and News Bulletin*, Vol. 13, No. 1.

Stotz, Lina, 2015, *Clean Clothes Campaign: An Overview of the Garment and Textile Industry in Pakistan*, Pakistan Country Report.

Tahir, S., T. Anwar, S. Aziz, K. Ahad, A.M. Mohammad and U.K. Baloch, 2001, "Determination of Pesticide Residues in Fruits and Vegetables of Islamabad", *Journal of Environment and Biology*, Vol. 22, No. 1, pp. 71–74.

Tariq, Muhammad Ilyas, Shahzad Afzal, Ishtiaq Hussain and Nargis Sultana, 2007, "Pesticides Exposure in Pakistan: A Review", *Environment International*, Vol. 33, pp. 1107–1122.

UNDP Annual Report Pakistan 2015 UNDP Islamabad, Pakistan.

Zia, Mummad Sharif, Muhammad Jamil Khan, Muhammad Qasim and Abdul Rahman, 2009, "Pesticide Residue in the Food Chain and Human Body Inside Pakistan", *Journal Chemical Society of Pakistan*, Vol. 31, No. 2.

5 Why 'education for all' does not turn into 'all for education'?

By 1854–55 there were 28,879 villages and at least as many schools in Punjab. In hundred years from now, literacy will be wiped out from Punjab.

—G. W. Leitner[1]

Two interesting quotations from the British Director of Public Instructions Punjab cited in the epigraph sum up the dilemma of the low level of literacy in contemporary Pakistan, choosing between the participatory and authoritarian path for mass literacy. The contention of this chapter is that literacy has been wiped out due to ill-advised policies of successive governments and can only be revived by bringing community back to centre stage. The context and dynamics of this ongoing tension between external interventions and community response are given below.

Policy framework

Education for All (EFA) in Pakistan is now a common goal of the government, civil society, the development community and international lending institutions. Article 37(b) and (c) of the Constitution of the Islamic Republic of Pakistan (1973) clearly declare that "the State shall remove illiteracy and provide free and compulsory secondary education within the minimum possible period; make technical and professional education generally available and higher education equally accessible by all on the basis of merit". Due to the 18th Constitutional Amendments made by the parliament in April 2010, education was made a provincial subject. Article 25-A of the Constitution of the Islamic Republic of Pakistan states: "State shall provide free and compulsory education to all children of the age of five to sixteen years in such manner as may be determined by law". The Inclusion of Right to Education in the constitution is an important signal of commitment to education by a democratically elected government. Taking steps to achieve universal literacy is also part of the Millennium Development Goals (MDGs) and Sustainable Development Goals (SDGs).

MDGs and SDGs: gazing at the stars and searching for the ground

Pakistan has made little progress in achieving the Millennium Development Goal (MDG) for education. It ranks among the world's worst performing countries in education. In 2012, 5.4 million children were out of school. A national household survey data shows that only 57 percent of children of primary schoolgoing age were in school by 2012–2013 (Malik: 2015: 3). The real disagreement is not on the goals but the means employed by various players to achieve the goals. It is important to mention here that SDGs and MDGs are global goals and they need to be indigenized in every country, administrative unit and settlement. The process of indigenization leads to engagement of all the relevant constituencies. It sets in motion the activities to document and present evidence on existing gaps and the available means to fill them. This process has not taken place either in Pakistan or anywhere else. One wonders how the path for achieving the goals can be charted without knowing the starting point (baseline), goal post and the resources available to cover the distance. Local communities and social entrepreneurs have done this work very effectively. Since IFIs, donors and UN agencies only deal with their borrowers and grantees, contributions of community outfits and social entrepreneurs are not taken into account by these agencies. A close review of actors outside the loan and grant circuit shows the effectiveness of entrepreneurial approach in indigenizing and achieving the global goals. There are many myths in circulation to explain underperformance of Pakistan in achieving MDGs and SDGs; these include lack of resources, lack of community 'awareness' and lack of 'technical expertise'. In the following sections, all these perceptions have been scrutinized in view of ground reality. While the authors of the MDGs and SDGs are gazing at the stars, they do not seem to have their feet on the ground.

A World Bank report, "Poverty in Pakistan, Vulnerabilities, Social Gaps, and Rural Dynamics", stated that the relative insulation of social spending from downward pressures during 1993–1998 was largely due to an infusion of $2 billion in support of the Social Action Programme (SAP). The report regretted that there was a serious problem of governance in Pakistan. Resources allocated to social spending over the past decade were largely used inefficiently and failed to have a significant impact on a dollar per dollar basis. Pakistan in fact exhibited persistent problems in most dimensions of governance that are relevant for sound public spending. The report added that there were leakages, "difficulties with bureaucratic structure and quality, weaknesses in the rule of law, and opacity in government decision-making" (World Bank: 2002: 6).

The budget conundrum

The Government of Pakistan and development professionals from the formal sector see the low level of literacy simply as an outcome of low budgetary allocations for education by the government. In view of continued failure of government in achieving the EFA objective, IFIs, UN agencies, international NGOs (INGOs) and

NGOs engaged in the advocacy for right-based development have kept raising this point with much greater force and made appeals for larger donor funding. In 2011, the Pakistan Education Task Force announced that Pakistan was facing an 'education emergency'. In view of the task force, this emergency was caused due to lack of funds for education. It is very important to analyze this claim dispassionately to clearly understand the nature of problem. For one thing, Pakistan's education budget of $7.5 billion is now almost equal to Pakistan's military budget ($8.2 billion). It does not depict a low level of financing or lower priority assigned to education. As aptly pointed out by Naviwala, "The problem in education is not of spending more but spending better". According to Naviwala, Pakistan is at the risk of overspending on education as its education budget far exceeds the UNESCO's recommendations that countries spend 15–20 percent of their budgets on education. She notes that the military budget is 2.9 percent of GDP and the education budget is 2.7 percent of GDP. If spending on education in the private sector is taken into account, Pakistan's spending on education crosses the 4 percent GDP mark deemed appropriate by donors and advocates of EFA. She has insightfully pointed out that private expenses on education are not taken into account because they are difficult to track (Naviwala: 2016: 9). Tahir Andarabi has also verified the contribution of low-cost primary schools. According to Andarabi, "there were more than 50,000 private schools with more than a third of total enrolment at the primary level" by 2005, and every village in Pakistan had at least one private school (Andarabi et al.: 2010: 1).[2]

It is also important to mention here that allocation of 4 percent GDP to education is possible under the 20 percent tax to GDP ratio. Pakistan's tax to GDP ratio is way below that mark (Vazquez and Cyan: 2015). As noted by the World Bank, "The tax-to-GDP ratio declined from 10.6 percent in 1999/2000 to 9.5 percent in 2011/12" (Ahmad et al. : 2014: 4). The Centre for Peace and Development Initiatives (CDPI) has also pointed out that "the tax to GDP ratio has never exceeded the 10.3% level achieved in 2002."[3] Therefore, any demand to increase allocations for education would not have much effect on resource mobilization. It is important to mention here that contrary to the issue of underfunding, government has mentioned underutilized capacity of primary schools as one of the key issues (GOP: 2014: 15). Pakistan has a primary education system consisting of 140,000 schools and 600,000 teachers. This capacity is not fully utilized to serve the children of schoolgoing age. It is also argued that the sector does not have an absorptive capacity beyond 1.8 percent of GDP (GOP: 2014: 23). In addition, Pakistan has a vibrant and well-spread private and civil society network of educational institutions. The Citizen Foundation, a CSO, runs one of the world's largest school networks in Pakistan. The private sector also provides far better education at almost half the cost to 40 percent of children in Pakistan (Naviwala: 2016: 1–2).

Increasing the education budget cannot help much because due to the political patronage of many primary schoolteachers, Pakistan has a large number of ghost teachers who either do not report to work regularly or don't attend school at all. Many teachers have been reported to be doing a second full-time job and drawing salary from school as absentee teachers. When the Government of Sindh introduced a biometric system to track down absentee teachers, many teachers had to travel back from Dubai and London to claim their salaries. It is important to note here that government teachers are paid very well, drawing five times the salary of private schoolteachers. The low level of enrolment therefore cannot be explained due to lower fiscal incentives for teachers or lack of education facilities. The external funding sought to increase the enrolment rate constitutes a tiny part of the educational budget in Pakistan. Education departments cannot actually effectively spend the budget they already receive. Provincial education departments are not able to spend more than 50 percent of the budget they receive as a development budget for constructing new schools. This happens because budgetary allocations for education are not based on planning and a broken budgetary process does not function to connect needs with resources (Naviwala: 2016: 13). DFID and the World Bank, two of the largest supporters of education reforms in Punjab, share the hunch that their funding is fungible, and gains made with their support over the last decade have been marginal and led to only a 1 percent increase in enrolment (Naviwala: 2016: 17).

Tim Unwin, a former employee of DFID in Pakistan, noted during his recent visit to Pakistan that

> There was also a strong perception that those involved in the design of the project had not grasped the actual realities of the educational challenges on the ground in Punjab. The truth of this is much more difficult to judge, but there was undoubtedly a feeling that the views of influential "outsiders", who rarely visited schools and villages on the ground, but spent most of their time talking with senior government officials in offices in Lahore or Islamabad, had been prominent in shaping the programme.
>
> (Unwin: 2016: 3)

Naviwala has mentioned former US Ambassador to Pakistan Cameron Munter, writing in a blog for the Brookings Institution that "We measured our commitment to Pakistan by how much we spent rather than assessing our impact" (Naviwala: 2016: 29).

The budget allocation for education is mostly utilized for current expenses, that is teacher's salary which averages to 89 percent at the national level, leaving a meagre 11 percent to raise the quality of education. The same pattern is repeated at the provincial level (GOP: 2014: 17). This reinforces the view that spending better is more critical than spending more in improving the quality and outreach of education facilities.

Tradition revisited: education for all; barriers and bridges

Naviwala and Haris Gazdar emphasize that elites matter in establishing vibrant educational systems. They have perhaps overlooked that communities also matter – and matter much more. In this respect it is important to mention that the 'good Samaritans' of the colonial elite played havoc with the glorious community education system in the areas which now constitute Pakistan through their policy biases for 'modern education'. After the British annexation of Punjab, the colonial government encroached upon the community space for provision of universal basic education and replaced it with the thinly spread government schools in the name of 'modern education'. It led to erosion of universal literacy from Punjab. Showing his dissent to the elite-based educational system, British Director of Education in Punjab G.W. Leitner had predicted that a hundred years from now literacy would be wiped out from Punjab.[4] Contrary to the elitist perception that communities across Pakistan are averse to education, these communities continue to support educational efforts with their generous contributions. Half of Pakistan's indigenous philanthropy goes for madrassa education, which is almost equal to the size of the national budget in the social sector (Andrew: 2013: 1; Bonbright and Azfar: 2002: 58–59).[5] Tens of thousands of children learn on-the-job technical skills through apprenticeship programmes. The colonial bias against universal education found continuity through 'market-based' policy choice of Pakistan's early planners who decided to leave the matter of human development at the mercy of 'market forces'.[6]

Large swathes of areas that now constitute Pakistan had in place a very sound and firm tradition of providing education based on self-help by beneficiary communities a little more than a hundred years ago. This self-help system provided universal literacy to males and females in most of urban and rural settlements through a very elaborate management structure of traditional rural communities, called 'village corporation' or 'village republic' by many modern social scientists. The nature, level and outreach of this education was partly described by G.W. Leitner in his *Report on Indigenous Education in the Punjab*, published in 1882. This system of universal literacy based on the concepts of voluntary work and self-help was not known as NGO work and was much larger in scale compared to present NGO initiatives. It was built on the same conceptual foundations on which educational activities have been successfully undertaken among low-income communities by NGOs during the past few decades. There were eight different types of schools, and the standard of education and pedagogical methods used in these schools were so high that they were imported in England in the earlier part of the 19th century. These schools included Pathshala schools, Chatshala schools, Gurmukhi schools, Sanskrit schools, Arabic schools, Persian schools, Quran schools and special schools for the merchant class.

Describing educational work in Punjab, Leitner stated that "By 1854–55 there were 28,879 villages and at least as many schools in Punjab. In towns, Delhi had 279 schools, Amritsar 143 schools, and Sialkot 38" (Leitner: 1882: 4). The rate of literacy was so high that in referring to a 1852 Settlement Report, Leitner mentioned, "In the backward district of Hushiarpur there was 1 school for every 19.65 males" (Ibid., pp. 2). Education was considered a basic moral responsibility of every educated individual, and in the words of Leitner, "I am not acquainted with any Native, Hindu, Mohammadan, or Sikh, who, if at all proficient in any branch of indigenous learning or service, does not consider it to be a proud duty to teach others" (Ibid., pp. 19). Continuing, he says, "Even among those educated natives who have not thrown aside social or religious restraints, I have known men devoting half of their slender incomes to maintaining schools or pupils at them" (Ibid., pp. 19). This voluntary spirit was integrally linked with a sound financial management system established by various artisan and farmer communities.

Demand for education and community support for this endeavour is corroborated by the experience of the CARE Foundation. CARE's founder Seema Aziz discovered during her visit to the flood-affected areas on the Ravi River in 1988 that a large number of children were following her wherever she visited. On her inquiry, she was told that "these children are idle and have no school to receive education". Ms. Aziz raised money from close friends and concerned citizens came together to help build a school. This school was named Care School. On the very first day, 250 children registered. Since then CARE has evolved to provide quality education to over 243,566 students in 866 schools across the nation; 833 of these schools are government schools managed by CARE (CARE: 2016: 6). The low level of enrolment and literacy has less to do with allocation of budget and more with mismanagement of resources and extremely low quality of education and its relevance with the job market.

Policy failure in the management of schools is reflected in the form of ghost schools, ghost teachers, the low quality of education and the dilapidated condition of physical facilities, hand pumps, classrooms and toilets. Some very good interventions proposed by formal sector professionals to improve the management of schools did not produce any results due to lack of preparation and participation of communities in the management process. This reinforces the point that community participation rather than donor funding is the centrepiece of universal literacy. Achieving the goal of universal literacy is linked to restoration of ghost schools and improvement in quality of education. Three distinct approaches were followed to deal with this problem: a formal sector 'participation' regime; community centered approaches based on the knowledge of community practices and informal schools; and the path taken by the social entrepreneurs. The comparison between these approaches is very illuminating.

Dealing with deficiencies: the case of ghost schools

There are an estimated 8,000 ghost schools in Pakistan where 40,000 teachers do not regularly attend classes but still receive a salary every month. Pakistan's former President Musharraf once said, "There are between 30 to 40,000 ghost schools, amounting to 20 percent of all schools."[7] In recent years, provincial governments have used various instruments to deal with the menace of ghost schools. These instruments include community watchdogs in the form of parents teachers associations (PTAs) and school management committees (SMCs); school voucher programmes; strict monitoring through monitoring committees and biometric systems; incremental development of demand-based schools and outsourcing of school management to social entrepreneurs. These interventions can be placed in two broad categories: authority-based formal sector approaches and opportunity-based informal sector and entrepreneurial approaches. An in-depth review of these approaches has shown that authority-based approaches have not succeeded in teaching new tricks to the old horses and produced dismal results. Entrepreneurial approaches have used existing gaps as entry points to make changes on the ground. Comparison of these approaches provides evidence that it is not the size but the management of resources that plays the key role in increasing access to primary education.

Community watchdog

To deal with the problem of ghost schools and teachers, federal and provincial governments introduced the concept of the SMCs and PTAs. SMCs and PTAs were supposed to watch the attendance of teachers, participate in school management and give feedback on quality of instruction. However, these committees were formed through official notifications; they did not receive any support and guidance to help them play their role effectively, and they were not given any fiscal powers to improve physical conditions of the schools and penalize absentee teachers. Review of these committees' performance revealed that they played a very limited role in school management and improving quality of education; they had an ambiguous role and limited funds; and due to a low level of literacy of their members and interference of elected representatives and bureaucrats, they operated with varying degrees of effectiveness (UNICEF: 2013: 58). A USAID report on the experience of SMCs and PTAs mentioned that administrative efforts by government institutions to engage these committees in the noble role to facilitate grassroots participation and build community ownership did not result in consultations with communities, as these committees were not sufficiently trained in their roles and responsibilities. It was only with systematic support by partner NGOs that these committees could assume an effective role (USAID: 2010: 2/30).

Zafar noted that

> a recent report on Punjab by Institute of Social and Policy Sciences (I-SAP: 2014) points out that SMC members were ill trained in procurement, school improvement plans, civil works, and construction to make effective use of Rs. 5 billion allocated to these committees during the financial year 2013/14.
>
> (Zafar: 2015: 16)

Strict monitoring

Strict monitoring of existing schools is another means used to end mismanagement of resources allocated for primary education. In the province of Khyber Pakhtunkhwa (KP), the government has decided to undertake strict monitoring of government schools to bring an end to leakages of resources. At present 25 percent of children go to private schools in KP and the government runs a voucher programme through their provincial education foundation, but the government's priority is to fix the public sector (Naviwala: 2016: 27). The chief minister of Punjab has taken a personal interest in enforcing strict monitoring of schools and made prompt decisions to fire ghost teachers and close down ghost facilities. This has created enormous pressure on the administrative staff in the education department, but the impressive short-term achievement can be transformed into long-term gains only by putting in place a decentralized, incentive-based system.

Awareness campaigns

Another strategy used by donors is to support 'awareness raising' on education. UK Aid has agreed to provide up to £20 million to a Pakistani NGO, Alif Ailaan, for this purpose. According to UK Aid, "This programme makes it more likely that we will succeed in making sure 4 million more children are in school, stay there, and learn more" (UK Aid: 2011: 3). It is important to mention here that this amount is sufficient to pay for high quality education of 5 million children – almost equal in number to out-of-school children. UK Aid review praised the programme as a "trailblazer in areas like education and school infrastructure, ... cross platform activism and engaging senior journalists and politicians" (DevTracker: 2014:). However, a DFID review of Teacher's Education Programme (TEP) commenting on the awareness campaign noted that

> Alif Ailaan has lots of activities but it is not immediately obvious to people who do not live and breathe the campaign how all these elements connect and work together to achieve the campaign's goals. This makes it harder to enlist partners because it is not clear to them how they fit into the bigger picture and where else they might add value.
>
> (DFID: 2013: 8)

Serious question about the value of this awareness programme arise when one considers the finding of Pakistan Economic Survey 2016–2017 that the country's literacy rate has dropped from 60 percent to 58 percent during this period. Speaking to the *Express Tribune* on International Literacy Day, observed on September 8, Alif Ailaan's campaign director Mosharraf Zaidi himself admitted that seriousness on the part of all stakeholders was needed. "Promises made in the Constitution or those made by the political parties of the country and non-profit organizations working for the cause had little impact on the situation on ground" (Sheikh: 2017).

Private–public partnership

School voucher programmes

In addition to SMCs and PTAs, a recent demand-based approach taken by the Government of Punjab is to strengthen the role of community members by providing them school vouchers. The government is not using resources to build new schools. Being cognizant of the presence of low-cost primary schools in low-income areas, especially rural areas, the government issues a monthly voucher worth US$7 for every child of schoolgoing age to enable parents to send their children to the school which they think provides quality education. This reduces the cost of educating a child to less than half the per-child cost of government schools. In this way, Punjab is also utilizing existing school infrastructure. Due to this intervention, half of Punjab's publicly reported new enrolments in 2015 took place in private schools and 35,000 children switched from private to public schools. The provincial government of Sindh took a different approach and invited concerned individuals and groups to adopt schools under their Adopt-a-School programme (Naviwala: 2016: 27). The result is yet to be seen, but the acceptance of these approaches shows that downward accountability and entrepreneurial management of school system has drawn the attention of policy makers for making significant changes.

Demand-based schools

The case of society for community support for primary education in Balochistan (SCSPEB)

In 1990s Balochistan had 11 times as many boys' schools as girls' schools, and girls' enrolment was much lower than the national average. However, due to parents' willingness to send their girls to genderless schools, their enrolment tripled in a few years (Sperling et al.: 2016).[8] It dispelled the myth that Baloch and Pukhtun parents did not want to send their daughters to school. However, there was another piece of the puzzle. There were many abandoned schools closed on the pretext that parents did not want to enrol their daughters for education. One explanation of this anomaly is that schools were built in areas where a local resident contributed a plot of land to build the school, and in return the donor of land was given lifelong employment as a guard or a peon at the school. Schools served the purpose of providing a job to the land donor; nothing else. This practice was followed in other parts of Pakistan. Zeba Mahsud, an assistant education officer in the Taliban-dominated Waziristan Agency in the tribal area bordering KP, reported that there were 6,050 educational institutions in Federally Administered Tribal Areas (FATA) of Pakistan (Mahsud: 2006).[9] Out of these, 4,868 (2,905 for boys and 1,963 for girls) were functional, and 1,182 (683 for boys and 499 for girls) were non-functional for the same reason. The problem here has to do with the supply-driven management practices. The challenge was to find a demand-driven approach and put community in the driver's seat. Children's interest in education can be gauged from the pictures of schools given in Figures 5.1 and 5.2.

Figure 5.1 Picture of schools in Pakistan

Source: Courtesy Arado, *Express Tribune*, January 29, 2018

Figure 5.2 Picture of schoolchildren in Pakistan

Source: *Express Tribune*, July 17, 2016

At this point the Society for Community Support for Primary Education in Balochistan (SCPEB) stepped in. The SCPEB proposed that the process of setting up a school should start with the request for enrolment of female students in a settlement. When applications for a given number of students were received, parents of the applicants were asked to identify a local teacher. In remote areas, the requirements for qualifications of teachers could be relaxed to accommodate local teachers. For the first six-month probationary period, the community also had to provide a room for holding classes. This usually happened to be the teacher's home. This arrangement suited the teacher as well, because she could earn extra income without leaving her home. SCPEB's job was to monitor the regular functioning of the schools. Under SCSPEB's memorandum of understanding (MoU) with the Government of Balochistan, a school was regularized after SCPEB's verification that it had effectively functioned during the probation period. This example pioneered the concept of demand-based schools and led to the establishment of a large number of functioning schools in the province.

However, large-scale and holistic changes in demand-based education were introduced through two highly acclaimed social enterprises set up by two business houses in Pakistan to provide an equal quality of education to children of low-income communities. Both these initiatives wanted to let education play the role of the great equalizer. These two enterprises are the Citizen Foundation and the CARE Foundation. Under a private-public partnership with these and other philanthropic enterprises, the Punjab government and other provincial governments have handed over the management of a large number of government schools to the private sector and the Sindh government invites concerned individuals and groups to adopt schools under their Adopt-a-School programme (Naviwala: 2016: 27). These schools under private management have addressed the issue of enrolment and retention both through improvement in quality of education and better management of schools and upkeep of their physical facilities. Their superb performance is due to the excellence of leadership. It is such leadership, not the fiscal allocations alone, that can ensure education for all in Pakistan.

Private management of government schools: from supply-side to demand-side economics

Seema Aziz, co-founder of one of Pakistan's leading high-end fashion retail outlets, Bareeze, pioneered the practice of establishing genderless high-quality English medium schools in low-income rural and urban communities. Such an approach is unprecedented. In Pakistan, schools are segregated into English medium, Urdu medium and government and private schools. Private English medium schools are at the high end of the spectrum and provide the highest quality education. Urdu medium, government-run schools are at the lowest rung in terms of quality and retention rate. Therefore, provision of education itself does not create equal opportunities for the students. Seema Aziz believes that "Education is a great equalizer, but only an equal education can equalize". Her schools were developed in line with this vision. She stumbled on the idea of providing education to children while visiting a village on the outskirts of Lahore to provide relief items to flood-affected communities in 1988. She noticed that village children were following her around during the visit. When she enquired why all the children in the village were following her, she was told that there was no school in the area. She adds,

> I knew I had to build a school there. When I floated the idea, I met with opposition. Statistics on the hundreds of government school buildings standing empty were shown to me to prove the point that the poor did not want an education.

> (Aziz: 2012: 28)

The day her first school opened, 250 children lined up for admission – proving wrong the perception that the poor do not want to educate their children. She decided right in the beginning that her schools would admit both male and female students; her students would be taught the curriculum taught in upper-class English medium schools; and they would receive instructions from well-trained teachers. All the half-naked children with runny noses and matted hair were to be provided hygiene education and treated with love and affection. Teachers were considered the biggest asset of her schools. Aziz established the CARE Foundation (CF) to mobilize resources and expand her work to teach 1 million children. A picture of a CARE school (Figure 5.3) gives an idea of Aziz's concern for providing equal education opportunities as a great equalizer.

Figure 5.3 CARE school

As the news of her excellent work reached the government, she was approached to take over the management of 25 government schools in Lahore. She agreed to initially take over the management of 10 schools only. In Aziz's words:

> We pioneered a model of public-private partnership in the education sector. Through this model, we adopt government schools and run them. In the first step, we provide all the infrastructural support that the school requires, ensuring that the building is up to our standards and that there are toilets, clean drinking water, labs and a library available for the students. Most importantly, however, we put in our management system and teachers to supplement the existing teachers in the schools. Our management ensures proper monitoring and evaluation of the school to ensure that a standard of education is maintained. As a result, dropout rates fall, and children receive the education they deserve.
>
> (Aziz: 2012: 28)

The school adoption project started in 1998 in Lahore. Aziz agreed to take on the schools offered under the condition that she could monitor the education and train the teachers herself. After months of negotiations, they complied, and she adopted the first school on September 1, 1998 (Aziz: 2012: 28).

It is interesting to note that while the Government of Punjab gives a school voucher worth US$7 to children attending low-cost private schools, the CARE Foundation provides high-quality education at Rs. 450 rupees (US$5) to educate one child each month. That equals over Rs. 9 million (US$100,000) a month for her entire operation (Aziz: 2012: 31). She also felt that education was a non-perishable asset which could last after people lost other assets due to disaster or calamity. A key element of the CARE Foundation's strategy for making the change was by setting an example, not by lecturing or training. Aziz raised all the money privately; no money was contributed by government or donors. Elaborating her views on educational reforms, she rightly said "Countries don't belong to governments. Countries belong to the people."[10]

It is pertinent to note that compared to the performance of high- and low-end educational facilities in the private sector, around a quarter of children who reach grade 5 in government schools in rural Pakistan cannot read sentences – a task that they should have been able to achieve by grade 2 (ASER: 2015: 16). It should be enough to prove that it is not more spending but better spending that counts. Better spending has to be an investment, not a subsidy.

Investment in teachers is the key to universal provision for improved quality of education. According to Andarabi et al., teachers in government schools are better trained and educated but their performance is lower due to a disconnect between their performance and financial reward. Their salary is based on their education and seniority, not on their ability to produce results. Child centered learning and performance-based report cards would form important tools for connecting performance with reward. As brilliantly summed up by Andarabi, "The goal of policy in this environment is not the creation of a 'missing' market from the ground up, but to facilitate its functioning"

(Andarabi et al.: 2010: 3). The private sector can be supported through establishments of more female secondary schools in rural areas to enhance the supply of teachers, the provision of credit to the private sector to upgrade to secondary education, the provision of guidance on school management and improvement of syllabi and teacher training. On the demand side, school vouchers give parents the power to choose to send their children to a better school compared to PTAs or SMCs.

The Citizens Foundation (TCF), a charity established in 1995, has set up 1,060 schools with an enrolment of 165,000 children in urban slums and rural communities (TCF: 2015: 7). Their emphasis has been on providing quality education to the less privileged. Quality in case of CF as well as TCF has meant moral, spiritual and intellectual enlightenment. For the sake of better quality education, TCF have prepared their own textbooks and teacher guides and arranged tutorials for students who need extra attention. TCF's teachers' training is now supported by a digitized training programme, a library of e-learning modules and endowment building to ensure financial sustainability of their schools during the due course (TCF: 2015: 38). Efforts of high-end philanthropic initiatives by the social entrepreneurs are supplemented on a large scale by low-fee private primary schools. These private schools have re-appropriated the space which was encroached upon by the state during the colonial period.

Private primary schools

Private schools in Pakistan have filled the gaps left by the government schooling system. National Education Management Information System (NEMIS) data indicated that in 2012/13 there were 17,093 private primary schools in the country. This data perhaps does not take into account the existence of tens of thousands of low-cost private primary schools in low-income urban settlements and rural areas. According to NEMIS data, 4.8 million (34 percent) of children in the 5- to 9-year age group are enrolled in private schools. According to official data, 34 percent of boys and 33 percent of girls are enrolled in private schools (GOP: 2014: 6). Andarabi's estimates put the numbers in private schools to be much higher. Andarabi also notes that quality and price are two most important factors determining the school enrolment and retention ratio, and private schools outperform government schools in this regard (Andarabi et al.: 2010: 1). He has noted that while education is free in government schools, the social cost of education through government schools is much higher due to the low quality of education imparted in the public sector. Low fees in private schools are explained by the presence of educated females in low-income communities, both urban and rural, who are willing to work at lower wages compared to men.

Informal sector

Informal educational institutions constitute the underbelly of the education system in Pakistan. They have covered largely the gap in school enrolment in the public sector. The education facilities created by the informal sector are poor communities' response to the unmet demand for education. These institutions include non-formal basic education centres, technical training centres for on-the-job learning though apprenticeship (*ustad-shagird*) system and madrassas. The solution of formal sector professionals for dealing with low-quality skill training and education in informal enterprises is to look down upon training in these institutions and seek elimination of this system, not improvement. They have not proposed their upgradation and integration with the formal system of education. Formal sector experts have strived to eliminate and replace them due to their limited understanding of the strategic part played by the informal sector to educate the poor children. These facilities are blemished for violating child rights, producing terrorists and imparting low-quality education without taking into consideration the ground reality. Their complete integration in the education system is the key challenge in the way of providing universal primary education.

Non-formal basic education

In response to ghost schools, many community-based organizations and NGOs established non-formal schools in the form of street schools and home schools in the 1980s and 1990s. According to government estimates, at least 2.5 million students are enrolled in more than 13,000 such schools across Pakistan. Local female teachers in their homes run more than 80 percent of these schools. At the end of grade 5, students are given the option to appear in formal sector examination and continue to grade 6 in formal schools on passing the examination (GOP: 2014: 5).

Technical education

Technical and vocational education for the development of skilled workforce and middle-level technicians comprises three years of education after secondary level (grade 10) or short certificate courses in various skills training for boys and girls ranging from 6 to 8 months (Ansari and Wu: 2013: 1). Major issues faced in technical education relate to relevancy, access, quality and equity. While the majority of the workforce is illiterate, government-run skill training institutions require 8–10 years of schooling as the entry requirement. The cost of formal skill training is another barrier to entry, as the average private cost of training at vocational and technical institutions is Rs. 19,951 for males and a little less for females. It is important to note here that the informal sector employs 78.8 percent of youth (ages 15–24) and 68.4 percent of those aged 25 or older outside of the agriculture sector (Janjua and Naveed: 2009: 1).

> The system is well suited to the needs of poor families in the sense that apprentices are provided small grants/wages during training. In addition, informal skill acquisition is important for a large number of individuals informally engaged with formal enterprises.
>
> (Janjua and Naveed: 2009: 3)

Informal skill training is confined to low-end businesses and is characterized by low productivity due to complete lack of support from public and private formal training systems.

Due to these characteristics, the number of youth acquiring skills through an apprenticeship programme (ustad-shagird system) is twice the number of students receiving skills training in government institutions. Against the estimated 63 million entrants in the labour market, formal sector institutions were expected to enrol 238,687 students in 2005, and only 1 percent of the labour force has received any technical and vocational training (Janjua: 2011: 99). The informal sector, on the other hand, employs 3.3 million school-aged children. These children are engaged in diversified occupations, including loading and unloading of goods, hotels and restaurants, fishing, auto workshops, rag picking, shoe shining, begging, glass bangle making, surgical industry, tanneries, fishing and domestic help. They are paid very low wages and, in many cases, given advances which leads to debt bondage of their families (LWP: 2011: 1). According to ILO estimates, 1.3 million children ranging from 15 to 17 years old were at work in hazardous employment in such enterprises (Khan and Lyon: 2015: 71).

Only 3 percent of all students enrol in skills training programmes, and government has played a little role in this sector. Government training institutions suffer from low employability, a high dropout rate and a weak link with the market demand. The formal sector also gives preference to on-the-job training. The relevance of technical and vocational education (TVE) curriculum provided at government institutions is perceived to be reasonable by 27.3 percent industrial employers, and 36.4 percent see it as relevant. This happens due to little involvement of industry in curriculum development (GOP: 2009: 37). The enrolment in TVE institutions is 105,000, which corresponds to only 1.4 percent of 14- to 15-year-olds, with another 115,000 engaged in tertiary-level diploma and certificate programmes, and over 326,000 enrolled in bachelor's degree programmes (GOP: 2009: 57–58).

Madrassas

There are 13,240 religious schools in Pakistan offering free religious education with free boarding and lodging to students, who mostly hail from destitute families. These madrassas are usually managed by local communities and are financed through charity and donations (GOP: 2014: 4). They are built incrementally, have a transparent financial management system and an inbuilt downward accountability regime. There is zero dropout rate in these schools. These schools receive 94 percent of the resources contributed by individual philanthropists for community services. Many of these schools teach curriculum taught in government schools to their students as well. Contrary to common perception, some of the madrassas teach subjects such as Urdu and English languages, mathematics and general science in addition to religious subjects. There are five federations of religious schools and all of them have agreed in principle to integrate government curriculum in their syllabi. Formal sector professionals oppose supporting madrassas on the ground that madrassas breed militancy.

It is important to mention here that the militancy associated with madrassas started taking root in 1980 when textbooks developed by the University of Nebraska–Omaha and published by USAID were made part of the text to motivate youth for 'holy war' against the Soviet Union. With state patronage, the number of these madrassas increased to 10,000 and enrolment to more than a million students. These madrassas were mainly funded by the United States and Saudi Arabia (Ahmed: 2009: 3). The total number of students going to madrassas is less than 1 percent of all students (Bergen and Pandey: 2005: 1). Studies exploring the linkage between terrorism and madrassa education have found engagement of a minority of madrassas (10–15 percent) with militant politics. It is also worth mentioning that this trend began at tandem with radicalization of textbooks in public schools (Ahmed: 2009: 5). The government's effort to reform madrassa education and integrate it with the secular education system under Pakistan Madrasa Education Board Ordinance of 2001 has not been received well by madrassas due to a low level of trust between the government and madrassa establishments.

Washington Post investigators reported that the United States spent millions of dollars producing fanatical textbooks:

> The primers, which were filled with talk of jihad and featured drawings of guns, bullets, soldiers and mines, have served since then i.e., since the violent destruction of the Afghan secular government in the early 1990s, as the Afghan school system's core curriculum. Even the Taliban used the American-produced books. . . . One page from the texts of that period shows a resistance fighter with a bandolier and a Kalashnikov slung from his shoulder. The soldier's head is missing. Above the soldier is a verse from the Koran.
>
> (Stephens and Ottaway: 2002)

Defending the decision to supply books with fanatic religious contents for school education, USAID spokesperson Kathryn Stratos said, "It's not AID's policy to support religious instruction. But we went ahead with this project because the primary purpose . . . is to educate children, which is predominantly a secular activity" (Stephens and Ottaway: 2002). What a brilliant attempt to justify hate-filled 'religious' teachings as secular.

Government is dragging its feet in integrating these schools into the formal education system like other sections of the informal economy. There are 1.79 million students enrolled in the religious schools at all levels; 1.1 million of these are boys and 0.69 million are girls. Around 58,000 teachers are employed in these institutions, 13,000 of them female. These figures should be enough to blow away the perception that religious schools are against female education (GOP: 2014: 6). The same is true about their connection to terrorism. Destruction of a small number of schools by a small minority of students educated at madrassas does not vouch for the majority of madrassas. However, military training received by these militants is not at all provided at madrassas. It is ironic that Western media has constructed a 'terrorist' image of madrassas and constructed counter-terrorist heroes for political expediency. It has not helped the cause of education at all. Trying to fabricate public consent based on a mis-representation of reality amounts to an insult to people's intelligence.

A comment of Seema Aziz in one of her interviews is very revealing in this regard. During her interview with Harvard Oral History Archives Professor Tarun Khanna, Khanna asked her, "What did you think when Malala [Yousafzai] got the Nobel Peace Prize? What thoughts went through your mind?" Aziz replied:

> I'd say surprised . . . everyone's trying to say the Taliban has some effect on literacy rates in Pakistan. They don't. The Taliban might have an impact on less than 1 percent of the area in Pakistan – I haven't noticed any visible impact she's had on education in Pakistan. That school she attended is still running, just like it was before. Everyone was going to school in that area. Her friends are still going to school. Nothing really changed.
>
> (Khanna: 2016)

What Aziz did not say was that the conferral of award was political; so has been the official assistance for improvement of education. There is nothing wrong with being political, but actions based on political expediency from the University of Nebraska project to the conferral of the Nobel Prize have insulted the intelligence of people. It has widened the gulf between the donors and public opinion. What Pakistan needs for promoting universal literacy is trust building between various partners more than anything else.

Where does the buck stop?

Achieving the target of quality universal education is not about demanding rights or seeking funds for globally set targets; it is about building bridges on the ground. A review of the ground situation reveals that government-run schools incur high costs and provide low-quality education. This is due to a deficient management system caused by ineffective monitoring, lack of a performance-based assessment system and rationalization of incentives. The informal sector provides low-quality education at a low cost. These schools cater to the needs of millions of schoolgoing age children. According to government estimates, at least 2.5 million students have been enrolled in more than 13,000 non-formal schools across Pakistan. The informal sector also employs 3.3 million children of schoolgoing age. There are 13,240 religious schools in Pakistan. There are 1.79 million students enrolled in the religious schools at all levels; 1.1 million of these are boys and 0.69 million are girls. Improvement and upgradation of these schools can lead to their integration in the mainstream and help in a significant way to achieve the target of education for all.

Private schools in low-income rural and urban communities provide good-quality education on a much lower budget. Their main constraint is finding well-educated local schoolteachers, and they can produce these teachers by upgrading their schools to a high level. With provision of credit from the government, this transition can take place much more rapidly than pouring money on 'awareness raising' or 'capacity building'. Social enterprises have set up the best practices in providing education to out-of-school children in low-income communities and urban and rural slums. They provide high-quality education at no or low cost. They have also been able to enrol, retain and educate students in the millions, enabling some of them to rise high in their career paths. Due to their impressive performance, various provincial governments have handed over the management of hundreds of schools to these entrepreneurs.

The CARE Foundation, for example, has evolved to provide quality education to over 243,566 students in 866 schools across the nation; 833 of these schools are government schools managed by CARE. CARE's cost of education per child is US$5 – much less than the value of a school voucher. The Citizens Foundation, a charity established in 1995, has set up 1,060 schools with an enrolment of 165,000 children in urban slums and rural communities. The schools run by these charities are supported by donations from community members. They do not receive any donor funding. To make their initiatives sustainable they are also building endowments. Creating endowments for providing sustainable universal education is a revival of a sound tradition that was disrupted during the colonial period. Local endowments and better spending are the key words for turning around the dismal situation in the education sector.

The answer to low achievement in basic literacy is better spending, not more spending alone. A summary of how this money may be better spent is given in Table 5.1. Critical importance of better spending can be appreciated by looking at the past use of donor assistance in Pakistan. As William Easterly points out:

> Pakistan received $58 billion in foreign aid from 1950–99. However, it systematically under-performed on most of the social and political indicators. Pakistan was the third largest recipient of ODA after India and Egypt during

Table 5.1 Policy options for universal quality education

Type of School	Quality of Education	Required Intervention
Government schools	High cost, low quality	Monitoring Performance-based assessment Rationalization of incentives
Private schools	Low cost, good quality	Provision of credit to build high schools Regulation School vouchers
Social enterprises	Low cost, high quality	Public-private partnership
Informal sector	Low cost, low quality	Regulation and guidance Provision of credit Integration

1960–98. If it had invested all the ODA during this period at a real rate of 6 percent it would have a stock of assets equal to $239 billion in 1998, many times the current external debt.

(Easterly: 2001: 3)

This view is reinforced if we consider that the gains made with the multimillion-dollar funding from DFID and the World Bank over the last decade have led to only a 1 percent increase in enrolment. It is also not certain if this funding was used for increasing enrolment or not. As informally mentioned by the donors, it was fudged funding; it was provided in response to reporting on increase in the numbers of students enrolled, not on the money spent by the government for this purpose.

Notes

1 See G. W. Leitner (1882). *Indigenous Education in Punjab During the British Period,* Calcutta 1882, Reprint Delhi: Amar Prakash, p. 4. It is important to note here that Punjab at that time spread over the area from Peshawar to Delhi and included most part of the areas now in Pakistan. Leitner captured the spirit of community devotion to the cause of education by saying that education was considered a basic moral responsibility of every educated individual. In his words, "I am not acquainted with any Native, Hindu, Mohammadan, or Sikh, who, if at all proficient in any branch of indigenous learning or service, does not consider it to be a proud duty to teach others." (Ibid: 19). He considered that British policies of eliminating local educational institutions on the pretext of "modernization" would play havoc with the system of universal literacy in Punjab. He therefore predicted that a hundred years from now literacy would be wiped out from Punjab. It is interesting that many experts blame the victims for the current low level of literacy in Pakistan and seek millions of dollars in foreign funding to create 'awareness' for the need to spread literacy.
2 These private schools seem to have replaced the community schools which were wiped out due to modernization policy of the British government more than a century ago. It clearly shows that lack of literacy is not due to lack of resources, lack of awareness or community resistance to education. As argued in the coming pages, a low level of literacy can be explained in terms of lack of support for these low-cost private schools.
3 Center for Peace and Development Initiatives (CDPI) Taxation System in Pakistan and its Impact on Children's lives, CRM Islamabad, n.d. CRM Islamabad.
4 The British government vehemently opposed the spread of its own 'modern' institutions of education through community efforts as well. Khan Abdul Ghaffar Khan (known also as 'Frontier Gandhi'), a pioneer of popular non-violent political resistance against the British Raj in the North Western Frontier Province (NWFP) and leader of the social welfare movement Khudai Khidmatgar (Men in Service of God), has repeatedly narrated in his autobiography how the chief commissioner of NWFP threatened him not to open schools in tribal areas and used police force to interrupt his visit to tribal communities. See Khan Abdul Ghaffar Khan's autobiography (Urdu edition) published by Hind Pocket Books Library, Delhi, 1969.
5 Bonbright and Azfar (2002) have noted that the impressiveness of the aggregate individual giving figure of Rs. 70.5 billion in 1998 is underscored when it is compared with government expenditures. The aggregate provincial and federal government spending on health and education in the 1996–97 budget was Rs. 84 billion.

6 Pakistan National Planning Board (1958). *The First Five-Year Plan (1955–60)*, Karachi.
7 See Christophe Jaffrelot (2004). *A History of Pakistan and its Origins*, Anthem Press, London, p. 187.
8 The myth of gender segregation in schools is quite recent. Even mosque schools have traditionally had genderless schools in existence for centuries. It is mentioned in the Punjabi folk tale, *Mirza Sahiban*. Leitner mentioned that in places where girls did not attend schools, male members considered it their duty to teach females.
9 The memoirs of Zeba Mahsood, named *Paras* (which means Chemist's Stone in local language) contain a detailed account of the condition of functional and non-functional schools. She was a high school teacher from 1993–1997 in South Waziristan and later served as assistant education officer from 1999–2002. In March 2002, as a consequence of her report, 18 ghost schools were cancelled and 72 ghost teachers were terminated. She reported the ghost schools to the governor and risked her life for getting the salaries of ghost teachers stopped.
10 Interview with Seema Aziz, interviewed by Tarun Khanna, Boston, Massachusetts, May 5, 2016, Creating Emerging Markets Oral History Collection, Baker Library, Historical Collections, Harvard Business School.

References

Ahmed, Ather Maqsood and Robina Ather, 2014, *Study on Tax Expenditures in Pakistan: World Bank Policy Paper Series on Pakistan*, PAK 21/12, The World Bank, viewed from http://econ.worldbank.org.
Ahmed, Z. S. 2009, "Madrasa Education in the Pakistani Context: Challenges, Reforms and Future Directions", *Peace Prints: South Asian Journal of Peacebuilding*, Vol. 2, No. 1.
Andarabi, Tahir R., Jishnu Das and Asim Ijaz Khwaja, 2010, *Education Policy in Pakistan: A Framework for Reform, IGC Pakistan*, Policy Brief.
Andrew, M., 2013, "The Generosity Graph: Pakistan and the World Giving Index 2012", *Dawn*, Website.
Ansari, Bushra and Xueping Wu, 2013, "Development of Pakistan's Technical and Vocational Education and Training (TVET): An Analysis of Skilling Pakistan Reforms", *Journal of Technical Education and Training (JTET)*, Vol. 5, No. 2.
ASER Pakistan, 2015, *Annual Status of Education Report Pakistan*, ASER, Islamabad.
Aziz, S., 2012, "Lighting a Path in Pakistan: Bareeze and CARE Foundation Pakistan", *Innovations Case Narrative – Innovations*, Vol. 7, No. 1.
Bergen, Peter and Swati Pandey, 2005, "The Madrassa Myth, Op-Ed Contributor", *The New York Times*.
Bonbright, David and Asad Azfar, 20002, *Philanthropy in Pakistan: A Report of the Initiative on Indigenous Philanthropy*, Aga Khan Foundation. Islamabad
CARE, 2016, *CARE Foundation Pakistan*, Organization Profile, Lahore. Online, viewed from www.carepakistan.org.
The Citizen Foundation (TCF), 2015, *Twenty Years of Believing in Pakistan: Annual Report*. Karachi.
DevTracker Project GB-1-202491, 2014, *Transforming Education in Pakistan*, Department for International Development, viewed from https://devtracker.dfid.gov.uk/projects/GB-COH-06407873-GB-1-202491 Project data last updated on 18/11/2014

DFID, 2013, *Annual Review Transforming Education in Pakistan (TEP)*, Date started: August 2011, Date review undertaken: July 3–5, 2013.

Easterly, W., 2001, *The Political Economy of Growth Without Development: A Case Study of Pakistan*, Development Research Group, World Bank.

Government of Pakistan (GOP) Ministry of Education, 2009, *Research Study on Technical and Vocational Education in Pakistan at Secondary Level*, UNESCO, Islamabad.

Government of Pakistan (GOP) Ministry of Education, Trainings and Standards in Higher Education, 2014, *Education for all 2015 National Review Report: Pakistan*, Academy of Educational Planning and Management Islamabad, Pakistan.

Khanna, Tarun, 2016, Interview with Seema Aziz, interviewed by Tarun Khanna, Boston, Massachusetts, May 5, 2016, *Creating Emerging Markets Oral History Collection*, Baker Library Historical Collections, Harvard Business School.

Janjua, Shehryar, 2011, "Is Skill Training a Good Investment for the Poor? The Evidence from Pakistan", *International Journal of Training Research*, Vol. 9, No. 1–2, pp. 95–109.

Janjua, Shehryar and Arif Naveed, 2009, *Skill Acquisition and the Significance of Informal Training System in Pakistan – Some Policy Implications*, Research Consortium on Educational Outcomes and Poverty (RECOUP) Policy Brief no.7.

Khan, Sherin R. and Scott Lyon, 2015, *Measuring Children's Work in South Asia: Perspectives from National Household Surveys*, International Labour Organization(ILO) DWT for South Asia and ILO Country Office for India, New Delhi.

Labour Watch Pakistan (LWP), 2011, *Child Labour in Pakistan*, viewed from http://labour watchpakistan.com/.

Leitner, G. W., 1882, *History of Indigenous Education in the Punjab since Annexation and in 1882*, Reprint Amar Prakash, Delhi.

Mahsud, Z., 2006, *Paras*, Peshawar, Pakistan.

Malik, R., 2015, *Financing Education in Pakistan: Opportunities for Action*. Country Case Study for the Oslo Summit on Education for Development.

Naviwala, N., 2016, *Pakistan Education Crisis: The Real Story*, Wilson Center, Washington, DC.

Sheikh, Ammar, *Express Tribune*, viewed from September 9, 2017.

Sperling, Gene B., Rebecca Winthrop and Christina Kwauk, 2016, *What Works in Girls' Education: Evidence for the World's Best Investment*, Brookings Institution Press, New York.

Stephens, Joe and David B. Ottaway, 2002, "From U.S., the ABC's Of Jihad", *Washington Post*, March 23 from https://www.washingtonpost.com/archive/politics/2002/03/23/from-us-the-abcs-of-jihad/d079075a-3ed3-4030-9a96-0d48f6355e54/?utm_term=.ae445ed164ac

UK Aid Project Summary, *Transforming Education in Pakistan 2011–2015 Business Case.*

UNICEF Pakistan, 2013, *Out-of-School Children in the Balochistan, Khyber Pakhtunkhwa, Punjab and Sindh Provinces of Pakistan*, Islamabad.

Unwin, T., 2016, *Education Reform in Pakistan: Rhetoric and Reality*, viewed on May 28, 2016, from https://unwin.wordpress.com/2016/05/28/education-reform-in-pakistan-rhetoric-and-reality.

USAID, 2010, *School Management Committees/Parent-Teacher Councils: Experiences in Capacity Building of Local Institutions and Their Contributions to Education in Earthquake-affected Pakistani Communities.*

Vazquez, Jorge Martinez and Musharraf Rasool Cyan, 2015, *The Role of Taxation in Pakistan's Revival*, Oxford University Press, Oxford.

World Bank, 2002, *Pakistan – Poverty Assessment: Poverty in Pakistan – Vulnerabilities, Social Caps, and Rural Dynamics.* World Bank, Washington, DC, viewed from http://documents.worldbank.org/curated/en/282761468775155766/Pakistan-Poverty-assessment-poverty-in-Pakistan-vulnerabilities-social-caps-and-rural-dynamics.

Zafar, F., 2015, *The Role of Donor Agencies in Education: Does It Pay?* Graduate Institute of Development Studies, Lahore School of Economics, Lahore

6 Universal health coverage; is health micro insurance the answer?

State of primary healthcare services in Pakistan

Basic health services to local communities in Pakistan were traditionally provided by health practitioners trained in a regime of disease diagnosis and treatment known as Greek medicine. Traditional practitioners also included Ayurvedic doctors trained in the traditional Indian school of medicine. Health workers providing pre-natal and post-natal care to pregnant women and providing childbirth and child health-related services as traditional birth attendants (TBAs) were known as *dais*. There was a wide range of paramedics attending to various healthcare needs of low-income people trained through a teacher-apprentice practice in Pakistan. Emphasis on preventive health and use of personal healing power was an important part of indigenous healthcare services. With the advent of British rule modern hospitals were set up, but they were mostly confined to large urban centres. Realizing the need for modern healthcare facilities to meet the needs of a rapidly growing population after independence, the government set up a three-tiered health services system, consisting of a Basic Health Unit (BHU) at village level, Mother and Child Centres and Rural Health Centres (RHCs) at the union council level and District Headquarter Hospitals at the district level. Below the district headquarter, tehsil hospitals were also established at some tehsil/subdivision headquarters (Amjad: 2009). In addition to paramedics, health workers, vaccinators and midwives are also part of this healthcare system. Most of the government healthcare facilities have purpose-built buildings, trained staff and a free supply of medicines but are conspicuous for having a large number of ghost facilities. Absentee doctors, nurses and paramedics and thefts of medicine are a common practice in many BHUs and RHCs. In case of missing facilities, low-income people either approach free clinics or faith healers and quacks.

The government's investment in building physical infrastructure for provision of healthcare services did not produce a commensurate impact. Pakistan has a range of low indicators for health. It falls among countries with an alarmingly high rate of infant mortality, maternal mortality and low life expectancy. Pakistan ranks 147th out of 188 countries and territories according to UNDP's Human Development Report 2016. Pakistan allocates less than 1 percent of GDP for health. Out of this extremely low budget, 80 percent goes to secondary and tertiary care services serving 15 percent of the population, and 15 percent goes to primary healthcare catering to the needs of 80 percent of the population. It is pertinent to note here that a large part of this budget goes to non-development funds in the form of salaries and other similar expenses. Only 33 percent of the population lives at most 5 kilometres from a health facility, despite the fact that there are 5,200 BHUs and 550 RHCs in the country. People in other areas have to travel long distances to receive outpatient care (Shaikh et al.: 2013). Gender bias in provision of services is reflected by the fact that in the first decade of the new millennium, 40 out of 100 rural health facilities had a sanctioned post for a lady doctor and only one out of three sanctioned posts was filled (Narjis and Nishtar: 2008).

Basic health services provided by the public sector are very deficient. The government has spent a great deal of money in building physical infrastructure for provision of health services but has a dismal record of managing the health facilities. The number of ghost doctors runs into the thousands, and the patient-to-doctor ratio is very high (Dawn: 2017). Theft of free medicines provided to government health facilities is very common. The government has expanded teaching facilities for training doctors to meet the increasing need for medical care, but a significant number of these doctors migrate overseas in search of greener pastures. The same happens in the case of paramedics. Due to low fiscal incentives, public sector doctors prefer to serve in urban areas or run private clinics in violation of their service rules as public servants.

The traditional form of community-based health insurance practice was known as *puchnee* in Punjab. Under this tradition, anyone visiting a patient would gift a certain amount of money to the patient to contribute to the cost of treatment. Such practices must have existed in other parts of Pakistan also, but they have eroded over time. Most common diseases among low-income communities in Pakistan are waterborne diseases including gastroenteritis, hepatitis, typhoid and skin and kidney infections. These diseases are preventable in nature. However, the provision of clean drinking water also suffers from underperformance due to mismanagement and the failure of a large number of government water supply schemes. No parameters for safe drinking water have been set by the government and a de facto privatization of safe drinking water supply has led to switching over to bottled water by the middle and upper classes and abandonment of the poor to consumption of contaminated water. One response to lack of adequate community health services emerged in the form of philanthropic healthcare facilities.

State of health and external assistance

There is no comprehensive data available on donor assistance received for improving primary healthcare services in Pakistan. Three points, however, sum up the nature of development assistance received by Pakistan. First, low health indicators were seen as an opportunity to sell loans by IFI to Pakistan, not to help improve the health indicators. These assistance programmes did not address the core issue of service delivery: management of resources (Altaf: 2011). Second, Pakistan's problem was not availability of resources but better use of resources, as Pakistan's health indicators were much lower than the indicators of other countries at the same income level (UNDP: 2016). Third, whatever development assistance was provided for human development, especially improvement of health indicators, did not lead to any improvement in health indicators at all. Most of the IFIs, development assistance partners and 'advocacy' NGOs have seen increasing the resource allocation as the most effective solution to the problem of healthcare mismanagement in the public sector (Altaf: 2011). Meanwhile, pioneers of knowledge-based approaches for improving healthcare services kept experimenting with innovative models to bring about effective change.

Donors have allocated significant amount of resources on training paramedics. But these paramedics also search for overseas jobs and migrate on the first available opportunity. Therefore, there is a perpetual shortage of health professionals, paramedics and healthcare services. As the poor cannot afford private healthcare facilities, in most cases they approach quacks or so-called faith healers, most of whom also fall in the category of quacks. Rationalization of financial incentives for medical and paramedic staff may offer an effective way for dealing with the phenomenon of ghost doctors and paramedics. Another option for dealing with the missing facilities is provision of universal health insurance coverage, giving poor patients the option to approach private clinics when needed. While government has shied away from putting in place a universal healthcare coverage despite the constitutional decree for making Pakistan an Islamic welfare state, some NGOs have taken the lead to walk the talk in this area.

Philanthropy in healthcare

Pakistani philanthropists responded to the paucity of healthcare services by providing free medical care. Free and subsidized healthcare services are being provided by both individual and family foundations, voluntary health services and service wings of some political parties for services ranging from basic healthcare to specialized treatment for eye diseases, cancer, leprosy and tuberculosis. Within the public sector, some leading medical practitioners have mobilized resources for providing free medical care to a large number of patients. One such expert is Dr. Adeeb Rizvi of the Sindh Institute of Urology and Transplantation (SIUT), who offers free treatment to kidney patients by mobilizing resources from philanthropists on a large scale. According to Dr. Rizvi, if you are serving a worthy cause you will never have a shortage of resources (Citation: 1998). A national level network for provision of a wide range of healthcare services is the Edhi Foundation founded by Abdus Sattar Edhi. Edhi ambulances would rush to pick up patients in critical condition, as well as victims of accident, violence and disaster before anyone from the private or public sectors would reach them. Edhi set up hammocks to receive unwanted infants, set up foster homes for mentally retarded children and shelter for women victims of violence. He buried the dead, collected and disposed of animal corpses and begged from the public to raise funds for the needy.

There are a large number of small, medium and large philanthropic healthcare facilities that cater to the needs of the deprived and low-income sections of the population. A German lady, Ruth Pfau, set up a leprosy centre in Karachi and kept serving this most neglected section of patients in Pakistan till her last breath. Specialized facilities have also been set up for treatment of cancer, eye diseases, blood transfusions and other ailments. There is no dearth of other similar initiatives. Business persons and families, artists, sportsmen and community leaders at various levels in the social hierarchy have set up their outpatient and healthcare units. Philanthropy, despite its wide coverage, does not offer a systemic response to meet the needs of people living below the poverty line. Many NGOs looking for a systemic response have tested various social enterprise models which might be expanded through self-replication or public sector extension based on field-level success. This search for alternatives started in the 1970s.

Social entrepreneurship – search for alternatives

Akhter Hameed Khan, a leading development practitioner in Pakistan, was of the view that NGOs should work as a research and development arm of the development sector and government as the extension mechanism. Public policy advocacy should be based on well-tested solutions in the field. Some community health practitioners shared this vision. Pioneering work in developing a voluntary healthcare model was done by the Sindh Graduates Association (SGA) in 1971. Another community health visionary was Dr. Nasim Ashraf, founding president of the US-based Association of Pakistani Physicians of North America (APPNA). Starting in the 1990s APPNA started experimenting with an integrated community health model for providing universal healthcare to the needy. Other alternative models for the provision of improved healthcare services by paid professionals were tested and developed by non-profit organizations as well as social entrepreneurs in the private and public sector. An important community health model was tested by the Health and Nutrition Development Society (HANDS) under the guidance of Dr. Ghaffar Billoo. With the beginning of the new millennium some experiments in provision of health micro insurance were also undertaken. A description of these experiments and the promises and perils of health micro insurance is given in the following sections.

Sindh Graduates Association (SGA) – the case for voluntary community service

SGA was formed in 1971 as Hala Graduates Association (HGA) by Suleman Shaikh, a doctor from Hala Town. Dr. Shaikh had graduated from a local high school many years ago and his once remarkably well-functioning alma mater was now in complete disrepair. He was one day approached by his school headmaster to help in restoring the school building. "What can I do?" Dr. Shaikh asked. "Visit the school that prepared you for your profession," said the headmaster. Dr. Shaikh visited and had the shock of his life when he looked at the school building. It triggered a process of thinking. Can the school graduates – doctors, engineers, civil servants, businessmen, professionals and wealthy – be asked to pay back? He started asking and discovered that people were willing to pay back, some in money but many in kind. The response was gradual and consistent and touched the hearts of so many. Eventually the Hala Graduates Association (HGA) was converted into the Sindh Graduates Association (SGA) and expanded all over Sindh. SGA members showed interest in improving health and education. This led to SGA's model for voluntary provision of education and health facilities. SGA helped in improving physical infrastructure of existing schools and setting up some new schools as well. Their healthcare model was simple: volunteer doctors were asked to donate as many days as they could for providing voluntary services. Their contribution could range from a day a week (mostly in case of general physicians) to a few days a year (for an eye specialist to hold an eye camp). Others could offer the use of their premises, donate land, construction materials or labour for setting up health facilities. Patients who could not be treated locally were referred to government hospitals at the district level. SGA subsequently started an ambulance facility for transporting patients to hospitals. Years later a female branch of SGA was set up headed by a female. In 1990s SGA started Friday clinics in Karachi and selected villages where patients could consult the doctor on payment of a nominal fee on a specific day in the week. SGA's model did not expand beyond Sindh but continues to serve the rural poor across Sindh. Their model is based on local supervision, operational autonomy and community participation.

APPNA SEHAT

APPNA's founding president Nasim Ashraf introduced a comprehensive health-care model known as APPNA SEHAT in 1990s. He along with a team of health-care experts conducted field visits to various areas of Pakistan for one and a half years to understand the community health issues and gather insights about effective local solutions. One of the key findings of the APPNA SEHAT model was that most of the diseases of the poor were waterborne diseases. These water-borne diseases accounted for 90 percent of ailments in low-income communi-ties. These diseases could be prevented by proper hygiene education, provision of safe drinking water, regular health monitoring and early detection of disease. This task could be provided by a team of male and female health workers and did not require the services of a trained paramedic. According to Dr. Hussain Bakhsh Kolachi, a health practitioner associated with a Tharparkar-based NGO, Banh Beli, basic ailments caused by consumption of unsafe water could be treated by a regime of seven medicines. These medicines could be imparted by a trained paramedic. This arrangement would not necessitate the presence of a doctor for a whole week in a rural clinic. However, the concept of local health service delivery based on curative and preventive services delivered by a doctor and a couple of community health workers was converted into a model by APPNA. This project was called APPNA SEHAT. The word SEHAT is an acronym for the project, but means 'health' in Urdu and other local languages (Khalil et al.: 2005).

Nasim Ashraf pioneered the idea of a community health programme at a few selected locations in Pakistan. Under this model, a block of 500 households con-stituted a viable unit to be covered by a team of one male senior health assistant

(SHA), two female health assistants (HA) trained by APPNA SEHAT (AS), two traditional birth attendants (TBA) and two male HAs. This team started work by paying individual visits to every household in the service area. They conducted a baseline survey and created a benchmark of disease profile in their unit. During their house-to-house visit, health workers also sensitized their clients about the need for provision of clean drinking water, hygienic practices, balanced nutrition and proper disposal of waste and the community's role in creating a healthy profile of the residents. Specific tasks undertaken by the team was to teach the mothers basic hygiene practices like washing hands before cooking and eating, preparation and dispensation of Oral Rehydration Solution (ORS) to infants suffering from diarrhoea and dehydration and the importance of vaccination and weight monitoring of children. Health monitoring cards were created for new born children and other members of the family. Health workers arranged for referral of the patients to local health units or hospitals. Community health education was an integral part of this programme. Community members were also mobilized to pool their resources to improve their drinking water supply system and eliminate sources of water contamination. These services were provided free of charge, as AS's investment in human development based on indigenous philanthropy.

Under this system AS provided health education, health record management and referral services; government health facilities provided curative treatment services; and patients paid out of pocket expenses for procurement of medicine. The community was motivated to invest in developing local clean drinking water schemes and contamination-free transmission of water to households. Once AS was sure that community members were able to continue the preventive health work without any external assistance, the AS unit was graduated to function independently. In the words of Dr. Hussain Bakhsh Kolachi, "when a community insider masters the skills of the outsider, the programme becomes sustainable." The AS programme was based on the same insight. This programme started in 1989 after a thorough study of existing healthcare practices in the public sector. It was named the village improvement model (VIM). The programme not only focused on community health but included special interventions for mothers and children. VIM was initially introduced in four units and during the course of the next two decades it expanded to 100 villages.[1] The total number served by AS might be much greater as they kept graduating out the units which were able to run without further assistance. AS also tried an 'adopt a village' approach to mobilize the Pakistani diaspora in the United States to finance its service provision in a larger area. The AS model offered a community-based alternative to government's ineffective service delivery system. It successfully demonstrated a participatory model for community healthcare and did not expand to a significant scale in rural and urban low-income communities (Khalil et. al: 2005). VIM was a good model that could be integrated with government's curative healthcare facilities for community health. Nasim Ashraf continued his work to develop an integrated community health model in the public sector under the National Commission for Human Development (NCHD). This model is known as the Gujrat model. A brief description of the Gujrat model is given in the following pages.

PRSP-Rahim Yar Khan (RYK) model

In the early 2000s, the Government of Punjab's advisor and philanthropist Jahangir Khan Tareen (JKT) conceived, developed and tested a model for dealing with doctors' absence from duty and regular provision of healthcare services at the RHC level in Lodhran district in partnership with the National Rural Support Programme (NRSP). Based on its considerable experience in rural development, the NRSP was able to provide technical support to develop and test the new model. The NRSP had organized rural communities in the district, and their village and women organizations could give a reliable feedback on the programme's effectiveness and benefit from its success. The model was expanded to Rahim Yar Khan district and named the Rahim Yar Khan (RYK) model. The NRSP's assumption was that due to low government salaries and limited opportunities for private practice to earn extra income, doctors were not attracted to serve in rural centres. Another reason for doctors' reluctance to serve in rural areas was lack of civic amenities and a vibrant social and cultural life, which would be amenable to relocation of their families at their duty stations. Young doctors were required to serve in rural areas for a number of years before seeking transfer to urban stations. They would therefore join a rural centre but never show up at work. Young doctors also did not have the means to purchase vehicles to pay daily visits to rural health facilities. This led to the emergence of a large number of ghost doctors.

Based on the success of Lodhran experiment, the Government of Punjab handed over management of 104 Basic Health Units (BHUs) to NRSP's sister organization, Punjab Rural Support Programme (PRSP), in April 2003. These BHUs formed clusters around RHCs and could benefit from the scheduled availability of the doctor in the nearest RHC. The RYK model tried to address the issues of absenteeism of doctors and paramedics by creating financial incentives for them. Under this approach, a doctor was handed over the charge of three rural health centres in close proximity to one another and had to visit each centre two days a week. Paramedics were supposed to take care of patients with minor illnesses through the week and patients needing special attention were referred to the doctor. The salary of each doctor was increased 2.5 times due to expansion of work to three health centres. Doctors were also provided interest-free loans to purchase vehicles to facilitate their mobility. The model worked very effectively and was replicated in various provinces under different names. In Punjab, the budget approved for PRSP's healthcare support is directly transferred by the district government to respective district PRSP Program Implementation Units (PIUs) for target health facilities. In addition, government provides funds for covering the PIU's operating costs.

The PRSP model also provided nominal performance-based incentives to junior staff. The model did not devolve management and provision of services to BHU level other than heath promotional activities at schools. For the purpose of community engagement, 'community support groups' were formed to interact with BHU management (Amjad: 2009: 6). The PRSP model succeeded in making a doctor available to patients from various BHUs on specific days and increased utilization of BHUs for curative services (Amjad: 2009: 14). Under this model women medical officers were also deputed in a number of RHCs. The programme was expanded to 11 districts after two months, and subsequently BHUs all over Pakistan were handed over to RSPs in each province under the President's Primary Healthcare Initiative (PPHIC).[2]

This model did not completely solve the problem of universal provision of healthcare to low-income communities. The RYK model had addressed the problem of availability of doctors and paramedics, but the RHCs suffered from serious management issues in providing proper care to the patients. To deal with healthcare management issues, another model, known as the Gujrat model, was tested to gain deeper insights to deal with the problem.

NCHD-Gujrat model

The Gujrat model aimed at improving access to primary healthcare through knowledge-based management. The model was designed and tested by the National Commission for Human Development (NCHD) headed by Dr. Nasim Ashraf (Amjad: 2009: 14).[3] The NCHD established a direct partnership with the district and provincial governments. The project was named 'Restructuring and Strengthening of Primary Healthcare System', and its aim was to align service delivery with the community's expressed need identified through a field-based survey. A door-to-door survey was carried out in the project area in 50 union councils of Gujrat for collecting baseline data on disease patterns, healthcare practices and issues of accessibility. An institutional innovation to ensure community access was creation of a community governance structure at the BHU level and the establishment of a local health council at the union council level. School health education was also an important component of the programme. The project undertook capacity building of BHUs, provision of equipment and capacity building of the staff and rationalization of the management system. In September 2007, NCHD's PHC model was expanded by the Government of Punjab for replication in 12 districts and 919 BHUs as the Punjab Integrated Primary Healthcare Model Programme (PIPHCMP).

Success of both the RYK and Gujrat models was based on a public-private partnership. These well-designed and tested models were based on sound research and close attention to management practices. Both programmes paid attention to rationalization of fiscal and operational management. Successful advocacy of these programmes by two highly renowned social entrepreneurs, Dr. Nasim Ashraf and Jahangir Khan Tareen, resulted in their expansion to the national level through the prime minister and the president.

Like the PRSP model, the NCHD model works at the BHU level and does not get involved in facility management, but it has a greater role in the integration of facility outreach and preventive services. It focuses on strengthening the referral system from the community level up to secondary-level hospitals. The PRSP model lacks such an arrangement, and this is one of the obstacles to its effective delivery of PHC services. Both models have strong political support and commitment from federal and provincial levels and therefore have been replicated at a rapid pace to other provinces. This rapid scale-up is a clear indication of the readiness of policy makers and politicians to support initiatives aimed at improving health services for the citizens (Amjad: 2009: 6). Success of these programmes highlights the fact that it is not the lack of resources but the lack of models based on sound management practices that explains the low quality of healthcare in the public sector. Such models can be developed through long-term engagement of experts in the field and not through short-term consultancies of experts not having deeper understanding of the existing practices.[4]

In the past two decades, Pakistan has made reasonable progress in building its primary healthcare (PHC) infrastructure, developing human resources for health and reducing rural-urban disparities in terms of access, coverage, and availability of services, but progress in various key health indicators is very slow (Amjad: 2009: 7). Altaf has shown that while migration of health professionals due to lack of fiscal incentives has been the main cause of deficiency in healthcare services, donors have been emphasizing capacity building of paramedics and evading the key issue of improved management of healthcare services (Altaf: 2011). During the course of testing public-private partnership models for access to primary healthcare, NRSP also considered the idea of developing health micro insurance as one of its microfinance products. This model held the promise of providing universal healthcare to the poor and increasing their choice to approach private service providers in the case of ghost facilities in the public sector. Such an arrangement could also exert pressure on government facilities by creating competition with the private sector. Most of all it could free the low-income families from the need to borrow or sell out family assets to receive primary healthcare.

Health micro insurance

NRSP – Adamjee Insurance Company partnership

Health micro insurance was a joint initiative of NRSP and Adamjee Insurance Company. NRSP had successfully organized thousands of communities in rural areas and administered its micro credit programme through its community organizations (COs). These COs received, disbursed and recovered the loans from individual borrowers and kept track of credit payments. This reduced the administrative cost of NRSP as a lending agency for serving small farmers. These communities could also serve as a mechanism for introducing and efficiently running a health micro insurance programme. The programme aimed at serving the bottom 20 percent living below the poverty line. This product line was not meant to generate any profit. If it could break even, then it could be expanded to a larger scale as part of Adamjee's corporate social responsibility portfolio. After intense discussions and careful calculations, it was decided to pilot test the programme. If it succeeded, it had an answer for universal health coverage to the poor.

NRSP approached the healthcare issue from the perspective of poverty alleviation. A World Bank report published in 2007 had indicated that 54 percent of low-income families were vulnerable because the economic shock caused by hospitalization expenses of one member of the family could push the entire family below the poverty line (RSPN: 2009: 16). This was verified by NRSP's field surveys as well. To cope with the financial shock caused by treatment expenses, low-income families were compelled to sell their land or livestock assets. In case of 40 percent of families, it took them at least three years to come out of poverty. The purpose of experimenting with health micro insurance products was to find a way to create an enabling environment rather than just improving some facilities or services. In early 2003 NRSP approached the Social Finance Programme of the International Labour Organization (ILO) to seek technical assistance for designing and implementing an insurance programme that could meet the requirements of rural communities in remote areas. An ILO expert suggested that NRSP should partner with an insurance company. NRSP invited Adamjee to design a product for its clients. The pilot phase of the programme started in October 2005 without any government subsidy.

Adamjee Insurance was a commercial organization and could not provide a permanent subsidy to execute the programme. A case for provision of subsidies to the low-income clients could be justified under corporate social responsibility, but the idea did not find traction with Adamjee's senior management. At this point NRSP offered its services to take over part of the administrative cost and administrative services to improve the cost-benefit ratio and make it attractive to Adamjee. The insurance policy offered was a group policy (Qamar et al.: 2007: 9). RSPs partnered with Adamjee Insurance and agreed to take over the claim processing responsibility on behalf of all the policy holders. Technically RSPN became the policy holder under the group policy issued by Adamjee and policy holders were considered legally insured persons (Qamar et al.: 2007: 17).

During the product designing phase, the first question was to decide the package: what is to be covered, who is to be covered and how much premium is to be charged? Next, procedures were to be decided: how soon is the claim to be submitted, what supporting documents are needed and who verifies the documents? Then there was the critical decision to be made regarding the mode of payment: whether insurance coverage will provide the facility for cashless transactions or reimburse the clients after clearance of claims. In that case claim reimbursement time was also of critical significance. In the case of cashless transactions, a network of insurance company–approved clinics was needed. The most important issue was the break-even scale of operation, as below a certain scale the insurance company would not buy in the product. These were no small challenges. It was decided from the outset that insurance will not cover outpatient treatment expenses.

The package designed consisted of the following items:

- Insurance would provide coverage for hospitalization, accidental death and disability of NRSP's partner community organization (CO) members. It did not cover pregnancy and delivery expenses of pregnant women.

- The package would cover up to Rs. 25,000 expenses for each of the items included in the package.
- The total premium of Rs. 250 per person per annum would be charged to every insurance holder.

Adamjee's share in premium was Rs. 208 and the RSP's share was Rs. 42 to meet operating costs. This average cost was determined on the basis of Adamjee's projection of claim amounts, supervision costs and a development surcharge to develop the product. NRSP's operating cost was calculated on the basis of additional staff that was required to maintain management information system (MIS), do some spot checks, track premium collections and do additional field work for processing the claims.

As a matter of practice, NRSP always tests a new product on a small scale to test its strengths and weaknesses and make required changes before launching the fully developed product on a large scale. Adamjee had targeted 500,000 clients in the first round, so NRSP could not test the product on a small scale. NRSP met the target by enlisting its borrowers as insurance clients and was able to register 185,000 policy buyers in year 1 and reached out to 4.5 million policy holders by year 3. Thus there were no financial constraints for Adamjee. Adamjee also got a cheque from NRSP before clients were even registered as policy holders. Selling health micro insurance policies on voluntary basis was a major challenge. NRSP made it mandatory for its borrowers to purchase a health micro insurance policy as part of their credit contract. The mandatory purchase of health micro insurance by micro credit clients solved the problem of scale but created difficulties in smooth functioning of the programme.

The first problem arose when policy renewal declined by 20 percent. NRSP's experience showed that accidental death of a breadwinner in a family created financial woes for the deceased member's family in addition to huge expenditures incurred on his treatment. Health micro insurance offered a solution by insuring repayment of outstanding debt in addition to hospitalization expenses of the policy holder (RSPN: 2009). NRSP decided to use it as a selling point to justify the health micro insurance premium as part of the credit issued to its members. NRSP convinced the borrowers that the payment of the premium for the micro health insurance policy will also cover their losses in case of death or disability of the borrowers. The insurance claim will help them pay back the loan and relieve them of unforeseen financial burdens (Qamar et al.: 2007: 23).[5] Scale achieved through mandatory insurance also helped in keeping the cost of collection lower than the cost of premium. This involved no collection cost for Adamjee. The programme was profitable for the insurance company because it had a 39 percent claim ratio and 50 percent loss ratio.

In deciding the premium and package, many other factors had to be taken into consideration. First and foremost, who should be the policy bearer: the individual, the family, the community organization or the RSP? It had administrative cost implications. If individuals were insured, it would enormously increase the administrative cost of the Adamjee Insurance Company and make micro insurance financially unattractive to them. Most of the target clients had to travel long distances to private or government facilities. If travel cost was added, it would raise the premium, but if it was not covered it would not give much relief to the policy bearer. It was estimated that cost of transportation to medical facility would be in the range of Rs. 3,000–5,000. Adamjee decided to include the travel cost in the cover. It was a reimbursement model which was very difficult in practice for cash-starved, rural, low-income clients. The policy originally claimed to reimburse the clients within 15 days, but in practice it took 3–6 months (Qamar et al.: 2007: 11). It caused serious issues with policy renewal. Gradually Adamjee introduced plastic cards and cashless transactions in many facilities in view of the cash-starved clients served by the scheme.[6]

There were two serious limitations of this model. First, NRSP offered health micro insurance to its micro credit recipients and did not cover the entire village community. Second, the premium was collected from the credit payments of the clients without their consent in many cases. NRSP had recommended only their micro credit borrowers for insurance because they could settle accounts with them in case of default by deducting the due amounts from their loan payment portfolio. This increased the chances of financial success of the programme but backfired as some of the clients refused to avail these services. Due to the limited number of policy holders and the high amount of treatment expenses claimed, the programme ran into serious management problems. Adamjee Insurance was reluctant in permanently subsidizing health micro insurance programme.

During the past 14 years NRSP has provided coverage to 2.9 million beneficiaries (including spouses) under the programme and an amount of Rs. 281 million has been reimbursed to the beneficiaries against hospitalization and death claims. Currently active members having insurance coverage are 600,000. The government recognized the health micro insurance programme as a social safety net and initiated the prime minister's national health programme to expand its coverage. The programme aims at serving 3.1 million poorest of the poor families (falling in the 0–32 band in the Poverty Score Card) of 30 districts in the first phase.

The challenges encountered by NRSP during the trial phase were (1) designing a product which meets the needs of clients in most comprehensive and inexpensive way; (2) delinking micro insurance with credit line and expanding the outreach; (3) integrating with technology to bring down the claim processing time and cost; (4) training staff and educating clients about the product package and claim requirement; (5) shifting from reimbursement to the clients to direct payment to the service providers; (5) improving the claim ratio (i.e. the percentage of premiums that are returned to policy holders as claim payments) from the current 37 percent upwards; as a general rule, the 75:20:5 ratio is considered to be a good business practice, in which case 75 percent claims are returned to policy holders, 20 percent goes for administrative cost and 5 percent is kept as profit for the insurance company (Qamar et al.: 2007: 27); (6) receiving complete documents from the hospital at the time of discharge so that no extra travel cost is incurred to claimant or claim is not lost due to lack of documentary evidence; (7) diversification to include affordable day care; and (8) bringing the people living below poverty line under the insurance cover.

From experimentation to expansion

Expansion through private sector innovations

Takafful

While public sector expanded the programme by subsidizing the premium of poor household and aiming to achieve the goal of universal healthcare for the poor, some innovative work was done in the private sector for this purpose. These innovative packages included Takafful, introduced under Islamic microfinance and Micro Ensure under mass marketing. Since Takafful was designed on the Islamic concept of brotherhood and sharing, under Takafful all premiums are deposited in the Takafful Fund which constitutes part of a Waqf Fund (Trust Fund) and surplus left over after the compensation of claims accrues to the same Waqf Fund. This surplus can be used to lower the service costs in future years (RSPN: 2009).

Micro Ensure

Some community health practitioners were of the view that provision of cross subsidy to the lowest 20 percent of population, whose premium payments were far below their treatment claims, could be provided through a mass marketing strategy for health insurance. This could be done by capturing the uninsured clients in the emerging markets. This strategy aimed to reach out to organized communities as well as middle income groups to generate large scale business and collect enough amount of premium to cover part of the services charges going to low-income policy holders. Mass marketing strategy aims at 80 percent of the potential clients who do not fall in the categories of the top 5 percent or the bottom 20 percent.[7] These clients provide the critical mass to make the programme successful. The idea was put into practice as Micro Ensure in 2001 and the programme was introduced in Pakistan in 2013. This business approach is an effort towards developing a model based on combining the financial interests and social responsibility considerations of the programme to render services in a sustainable manner. Micro Ensure has made another innovation. It builds on what already exists to reach out to the mass market. It works in partnership not only with microfinance institutions but also business organizations, especially banks and telecommunication companies with a large client base and instant access.

The programme claims to have 40 million customers in 20 countries.[8] Eighty-five percent of these customers had never had any insurance before. The strategy of capturing the emerging markets in the words of Micro Ensure is "rather than providing $1,000,000 coverage to 1,000 people as in traditional insurance, micro insurance provides $1,000 coverage to 1,000,000 people."[9] Their success consists in keeping the premium low, incorporating the services charges of existing illnesses in the premium and balancing their insurance claims with the premiums collected from the policy holders.

Micro Ensure has reached out to the mass market through a Lahore-based NGO, Damen; a telecommunication company, Telenor; and a host of other private enterprises. Damen's package covers a broad range of healthcare needs and is very user-friendly because it covers treatment from any registered hospital in Pakistan and pays Rs. 1,000 for each day of hospitalization in case of admission in a hospital due to illness or accident. The amount is doubled (i.e. Rs. 2,000) for any days spent in an intensive care unit (ICU) and a sum of Rs. 10,000 is paid in the case of childbirths through cesarean section. To ensure a quick and stress-free service, Micro Ensure ensures quick payment of claims (i.e. within three days of submission of documents). The claim payment ratio remained at 52 percent of the earned premium during July 2015 to June 2016.[10]

In partnership with a microfinance product, Easypaisa, and a telecommunication company, Telenor, Micro Ensure launched a product called 'Sehat Sahara', specifically designed to offer health insurance to the mass market in Pakistan. This provided a very convenient way for accessing healthcare for inpatient hospitalization and disability, offering a new innovative method of health financing to those in need. In the first year of launch, they enrolled 100,000 subscribers, that comes to 100,000 people (and their families). Another similar initiative was started by Oxford Policy Management (OPM). OPM provides technical assistance to the Kreditanstalt für Wiederaufbau (KFW)-funded Social Health Protection Initiative, which was initially aimed at insuring the poorest 20 percent of population in four districts of Khyber Pakhtunkhwa (KP) province and one in Gilgit district. It is now being upscaled to 60 percent of the population in all districts of KP, and a similar upscaling is underway in the Gilgit Baltistan area.

Expansion through philanthropy

With the exit of Adamjee Insurance, health micro insurance was taken over by Jubilee Insurance Company of Aga Khan Development Network (AKDN) as a social responsibility project. Jubilee made revisions in the insurance package and also agreed to subsidize the programme under its corporate social responsibility policy. Under the new package (1) the insurance premium was brought down to Rs. 200 from Rs. 250 and covered two persons; this meant covering one patient for Rs. 100; (2) coverage of pregnancy-related complications and childbirth was added; (3) the requirement for 24-hour admission in a hospital was waived; (4) the age limit for the insurance clients was enhanced from 18–55 to 18–65 years in December 2006; and (5) for claims, both cashless and reimbursement facilities were available for the policy holders. According to some estimates Jubilee only collects 25 percent in premiums against the value of claims under its health micro insurance programme. However, it is in line with the AKDN vision of serving the needy as a spiritual duty.

Expansion through the government: health micro insurance by the federal government

To protect the low-income families from falling below the poverty line due to health shocks, NRSP and two other RSPs, Sindh Rural Support Organization (SRSO) and Thardeep Rural Development Programme (TRDP), are providing health micro insurance to the destitute and the vulnerable households in eight districts of Sindh through an insurance company under the Sindh Union Council and Community Economic Strengthen Support (SUCCESS) programme. The insurance scheme is free of cost for the target households whereas financial assistance to meet the operational cost and the premium is provided by the European Union (EU). The scheme covers the nuclear family including parents. Each member of the family is covered for up to Rs. 25,000 per annum. Health micro insurance also provided support to the poorest households under the Union Council Poverty Reduction Programme (UCBRP) in Sindh by SRSO and Bacha Khan Poverty Reduction Programme in Khyber Pakhtunkhwa by Sarhad Rural Support Programme (SRSP) during 2009–2012. Initially the provincial government and donors provided financial assistance to meet the operational cost of the health micro insurance schemes and also paid the premium. The provincial government of Khyber Pakhtunkhwa with the financial assistance of KFW started a social health insurance programme in four districts – Mardan, Malakand, Chitral and Kohat – and now the programme has been expanded in the entire province by the provincial government through its own resources.

NRSP realized the need for a countrywide programme that could provide access for a large number of people to hospitalization services on a sustainable basis. For this purpose, NRSP initially carried out a dialogue with the provincial governments to pay premiums on behalf of the poor families. Policy makers in government were informed that providing health micro insurance is providing a social safety net to poor households who lie in the 0–18 poverty band.[11] Under Union Council-Based Poverty Reduction Programme (UCBPRP), the Government of Sindh introduced health micro insurance in 2009. In October 2009, Adamjee Insurance Company in collaboration with RSPs introduced a product of micro health family insurance for poor households covering married couples with children up to 18 years old and unmarried sisters. The service package with a ceiling of Rs. 25,000 per person per year includes visit to Out Patient Department (OPD), day care, diagnostic services, and hospitalization (for more than 24 hours' stay) and maternity care. The package also included accidental cover (disability compensation) and financial support of a sum of Rs. 25,000 as funeral charges in case of death of a breadwinner of the family (GOS: 2009–2012).

Health micro insurance was presented by NRSP to the federal government for implementation as well. The NRSP/Adamjee team was part of the task force of the Benazir Income Support Programme for Waseela Sehat (Healthcare Support).

Prime Minister National Health Insurance Programme had been in discussion with us to initiate a health micro insurance programme all over Pakistan.

NRSP calculated that if 40 percent of people living below poverty line could be covered at a premium of Rs. 40 per head it would cost the government only Rs. 6.4 billion (US$80 million). In terms of purchasing power parity GDP it amounted to 0.02 percent and in real terms it would only cost 0.06 percent of GDP making health for all possible within the meagre resources allocated for community health by the government.

(RSPN: 2009: 20)

Subsequently Prime Minister Muhammad Nawaz Sharif launched a state-run health insurance programme in 2015 and described it as the first step towards making Pakistan a welfare state. The scheme has been launched in Islamabad in the first phase and will be expanded to 23 districts in the second phase. Around 1.2 million families will get free healthcare facilities in the first phase. Currently, people who are living in 23 priority districts and earn less than US$2 a day (according to BISP Survey) are the beneficiaries of the Prime Minister's National Health programme.[12] A family can use up to Rs. 250,000 for priority healthcare services and Rs. 50,000 for secondary healthcare services. Family can benefit from the card till cash limit is available. Families can text their national identity card number to 8500 to check their eligibility in the programme. In case they have been declared eligible, they can receive their Pakistan Sehat card from the card distribution centre developed in their district. The cost of treatment after the patient has been admitted in the hospital shall be charged from the Pakistan Sehat card.

The programme is to be expanded to Punjab province with a $150 million loan from the Asian Development Bank (ADB). ADB will also provide a $430 million loan to expand social safety nets. The loan will support the Benazir Income Support Programme to expand the national cash transfer programme, which is paid quarterly to the mother in each poor household and will also be used for broadening and improving a pilot health insurance scheme and skills development programme (ADB: 2010, 2013).

Conclusion

A spectacular achievement of Health Micro Insurance (HMI) is that it has shown the path of universal healthcare coverage by the state by allocating only 0.06 percent of GDP for this purpose, much below the allocations demanded by advocacy NGOs. It also shows that primary healthcare programme can be solved by working under the system and living within the means. Jubilee Insurance of the Aga Khan Development Network (AKDN) has set a shining example for the civil society by demonstrating that civil society stands for achievement of public goals by private means. They consider it their spiritual goal. Their cost sharing is actually not a subsidy but investment in human development. This should have been the approach taken by the government as well, as the constitution declares Pakistan as an Islamic welfare state. Expansion of micro insurance by Takafful is also based on the concept of social welfare and community support. Micro Ensure translated the concept of corporate and community social responsibility into a market-based product. They used the concept of cross-subsidy to convert the idea of social responsibility into an insurance programme. What is surprising is that provincial and federal governments are trying to finance HMI not through their own budget but by international lending.

This borrowing for expanding the sound universal coverage programme will burden the poor in the guise of a welfare programme. It will result in payment of higher debt servicing and recovery of debt through indirect taxation unfairly hurting the poor. Questions that arise in this regard are as follows. Should the universal health insurance coverage for the poor be financed through bank borrowing or self-financing through higher budgetary allocation? Should the loan-based strategy be considered a strategy for electoral gains or a policy to serve the poor? Can we talk of a social welfare state without talking of social responsibility of the state and rich classes, and universal coverage through cross subsidy? ILO assistance for designing HMI was more in line with the practice of aligning poverty alleviation programmes with social reality of. the poor, but the ADB approach provides relief in the short run while adding to misery of the beneficiaries in the long run. The bottom line is that poverty alleviation is a moral question. Donors can provide the loan, but they cannot change the mindset by renting the conscience.

Notes

1 For latest information, visit www.janjonesworldwide.com/ashraf.html.
2 There are eight RSPs: Aga Khan Rural Support Programme (AKRSP), Sarhad Rural Support Programme (SRSP), Balochistan Rural Support Programme (BRSP), Punjab Rural Support Programme (PRSP), Sindh Rural Support Organization (SRSO), Thar Deep Development Programme (TRDP), Ghazi Barotha Taraqqiati Idara (GBTI), and National Rural Support Programme (NRSP), spread in five provinces and reaching out to 130 districts in Pakistan. The problem of scale cannot be solved with RSPs' partnership because NRSPs cover 5–10 percent of the rural population and only 33 percent of Pakistan's population lives within 5 kilometers' radius of existing BHUS.
3 National Commission for Human Development (NCHD) is a GONGO; an NGO established by the government. This status provided it power and prestige of a government body and autonomy and flexibility of a non-government organization. The first chairperson of NCHD, Mr. Nasim Ashraf, was given the status of minister of state by then President Pervez Musharraf, who held the status of patron. It was established by the Government of Pakistan in 2002 as a statutory autonomous federal body, mandated with the role to support and augment human development efforts in Pakistan. NCHD is registered under Ordinance No. 29 of 2002.
4 Outside consultants are no doubt very knowledgeable in their field, but without local knowledge they cannot contribute much in developing useful models. Dr. Akhter Hameed Khan used to call such technical assistance "the case of blind leading the blind". Samia Altaf has given a very lucid and funny account of one of the typical consultancy visit for providing technical assistance to Pakistan. The mission leader knows nothing about the health system in Pakistan, systematically evades the key issues and ends up repeating all the failed solutions as her recommendation. For further details see Altaf (2011).
5 Since Adamjee is a business organization, it does not share information on the claim ratio to help determine if the programme breaks even. The figure I have quoted was

calculated by the consultant who evaluated the programme. However, Adamjee's loss of continued interest in health micro insurance must have taken place due to the high claim ratio. In the words of Adamjee's former official, Ms. Saima Zafar, "Our health microinsurance (HMI) programme had been running with various RSPs (not alone with NRSP). During 2009–2012, we also started providing HMI programme with Government of Sindh and Khyber Pakhtunkhwa under Union Council Based Poverty Reduction Programme. After that government didn't continue the programme; which resulted in decrease in size of insured persons and premium amount. This had resulted negatively at our claims ratio. Therefore, Adamjee discontinued health microinsurance programme."

6 Telephone conversation with Karim Khan Qamar, November 27, 2017.
7 Telephone interview with Mr. Rehan Butt, country director of Micro Ensure Pakistan, October 26, 2017.
8 Micro Ensure website, https://microensure.com/, October 27, 2017.
9 Ibid.
10 Micro Ensure declined to provide any figures for their claim:administrative cost:profit ratio because under their contract with the partners they were obliged not to disclose any data.
11 The Government of Pakistan has compiled a Poverty Score Card (PSC) with the help of partner NGOs for most of the districts in Pakistan. It consists of 32 bands, zero being the lowest and 32 the highest. These PSC profiles are available on the website of National Database Registration Authority (NADRA).
12 The Benazir Income Support Programme (BISP) is a federal government programme for cash transfer to women living below the poverty line.

References

ADB Media Centre, *Pakistan's Punjab Gets $150 Million ADB Loan to Cut Infant, Maternal Deaths*, media release, viewed on June 24, 2010.

ADB Media Centre, *$430 Million ADB Loan to Help Pakistan Expand Social Safety Nets*, media release, viewed on October 24, 2013.

Altaf, S. W., 2011, *So Much Aid, So Little Development: Stories from Pakistan*, Woodrow Wilson Center Press/Baltimore Johns Hopkins University Press, Washington, DC.

Amjad, Dr. Sohail, 2009, *Review and Assessment of Various Primary Health Care Models in Pakistan*, USAID Pakistan.

Citation, 1998, *Citation on Syed Adib ul Hasan Rizvi, Ramon Magsaysay Award*, viewed from http://rmaward.asia/awardees/rizvi-syed-adibul-hasan/ 1998

Government of Sindh (GOS), *Programme Report-year 2009–2012*, viewed from www.ucbprp.net.pk/publication.htm.

Khalil, Moazzam, Mohammad Ayub, Farooq Naeem, Mohammad Irfan, Shakir-ur-Rehman, Shakir Ullah Bacha, Muhammad Babar Alam and Nasim Ashraf, (2005)., "APPNA SEHAT: A Description of the Development of a Health Education Programme in Rural Pakistan", *International Journal of Health Promotion and Education*, Vol. 43, No. 4, p. 137.

Mansoor, Hasan, 2013, *Health Dept Inquiry Finds over 3,500 'ghost doctors'*, viewed from https://www.dawn.com/news/1050245.

Narjis, R. and S. Nishtar, 2008, "Pakistan's Health Policy: Appropriateness and Relevance to Women's Health Needs", *Health Policy*, Vol. 88, No. 2–3, pp. 269–281.

Qamar, Karim Khan et al., 2007, *The Beginning of Health Microinsurance in Pakistan: A Review of RSPN-Adamjee Insurance Scheme*, RSPN, Islamabad.

Rizvi, "Pakistan's 'miracle' Doctor Inspired by NHS", *BBC News*, viewed from www.bbc.com/news/world-asia-29648454.

RSPN, 2009, *Proceedings of the First Micro Insurance Conference*, the Rural Support Programmes Network (RSPN), Pakistan.

Shaikh, S. B., E. I. Ejaz, A. D. Achakzai and S. Y. Shafiq, 2013, "Political and Economic Unfairness in Health System of Pakistan: A Hope with the Recent Reforms", *Journal of Ayub Medical College Abbottabad*, Vol. 25, No. 1–2, pp. 198–203.

UNDP, *Human Development Report 2016*, viewed from http://hdr.undp.org/sites/default/files/2016_human_development_report.pdf.

7 Poverty alleviation and arithmetic of the poor

Is civil society an appropriate receiving mechanism for poverty alleviation?

This chapter will conclude with a closer view of frugal science initiatives based on the arithmetic of the poor in relation to poverty alleviation, social change and economic development. Frugal sciences depict the ideas, practices and initiatives regarding poverty alleviation based on the reality of low-income communities, but the reality of the poor is perceived differently by donors and advocacy NGOs. These three distinct perceptions have different implications on poverty alleviation, aid effectiveness and sustainable development. Aid can be effective only if it is channelized through the most appropriate receiving mechanism. This view is based on the critical evaluation of three distinct mechanisms: (1) NGOs working on IFI-inspired initiatives as donor contractors; (2) rights activists demanding compliance of the state with international obligations; and (3) NGOs building their work based on the arithmetic of the poor and conceptualizing solutions based on the social reality of the poor and negotiating for their implementation. The pluralist world of the civil society creates the hope that through selection of the most appropriate practices and institutions, aid may help change for the better (Iqbal et al. 2004; AKF: 2003; EDC: 1996; Habib: 2002; Hasan: 2007; UNDP: 1991). This chapter will present the perspective and trace the origin, development and dynamic interaction of these tendencies: from the age of the 'village republic', the name selected by the British for the legendary self-sufficient village in India (now South Asia), to the global village in shaping up the vision, structure, and impact of civil society in defining the citizen-state relationship for poverty alleviation. The scale of indigenous CSOs' success might be small, but the significance of their approach is far more important. Insights gained from the work of indigenous NGOs are invaluable. As the saying goes, "A wise eye can see an ocean in a drop of water."

Can development be exported?

Pakistan presents a special case of foreign assistance for economic development and poverty alleviation. Pakistan was a close ally of the United States and the NATO bloc during the Cold War period, and therefore periods of rise and decline of foreign assistance coincided with the level of geopolitical interest of Western allies in the region (Naved and Nabi: 1991). It seems that political expediency more than economic wisdom played an important role in the provision of foreign economic assistance to Pakistan.[1] For example studies by experts have revealed the existence of negative relationship between tax revenue collection and saving on the one hand and foreign aid on the other (Naved and Nabi: 1991). This weakens the case for economic assistance but fits in with the doctrine of 'development without tears', which was meant to appease the ruling elite to enlist its support against the Cold War enemies. However, this assistance cannot be considered a litany of failure. It had its share of successes and failures. Due to the price distorting impact of subsidized loans and tariff and quota restrictions, Pakistani industries were found to be adding negative value, which meant earning private profit in the face of social losses and increasing poverty in view of concentration of wealth. Due to protection against international competition, Pakistan's textile industry for example wiped out the rural handlooms industry instead of import substitution and added to the ranks of the unemployed. This started the process of disintegration of self-sufficient village economy known as the village republic and expedited the process of rural-urban migration. This strengthened rent-seeking behaviour among certain sections of the ruling elite.

Foreign assistance for poverty alleviation in the social sector has had mixed impact. Pakistan is home to a diverse group of civil society institutions: contractors of donor agencies, advocates of the 'universal' human rights agenda and visionary and idealist social entrepreneurs. While contractors work from project to project, 'advocacy' NGOs raise the voice without much success in social involvement; social entrepreneurs undertook initiatives and founded institutions that made impressive contributions to human development. These social entrepreneurs have largely relied on indigenous philanthropy and frugal human development models to realize their vision by working within the means and living within the system. Mostly they rejected any donor assistance because it compromised their independence. However, there are cases where social entrepreneurs accepted foreign funding, and they made the best use of foreign economic assistance. Review of aid in relation to contractors and social entrepreneurs provides us important insights about aid effectiveness.

In what follows, I have tried to identify different forms of development assistance and drawn lessons for a meaningful collaboration. A review of prominent cases of poverty alleviation depicts that selection of the right partner with knowledge of the ground reality, long-term engagement capacity and entrepreneurial vision played a crucial role in aid effectiveness (Ahmad: 2008; GOP: 2005). Aid was effective if it could become part of a programme where donor funding was used for a discrete activity, and the programme could continue its course without complete dependence on external funding. Aid also contributed to meaningful change where it was used for expansion of successful models (Husain: 1992). This way a time-bound, result-based intervention could become part of a process not restricted by the compulsions of fast-paced delivery. This type of aid could produce results if it was provided in response to a local need, not as an unsolicited supply-side intervention. Limited knowledge of the ground reality and ideological blinders of supply-side development assistance constrained its effectiveness.

Assumptions of donor-assisted development approach

Aid went wrong where assumptions of the donor were not in line with the ground reality. Most of the donor missions come in for a short duration and they have extremely limited knowledge of local conditions.[2] In the absence of local knowledge, consultancy missions consider that they are working on a clean slate and have to start from scratch. To them, existing local infrastructures, assets and capacity do not matter in the planning exercise. Consultant-driven development plans and projects propose solutions with similar factor intensity that prevails in the developed economies. They also assumed an accountable system of governance and a cash-based economy in place that could easily absorb funds for fast, large-scale and time-bound projects. This 'technical' viewpoint has its root in the theory and ideology of the free market.

Foreign economic assistance: theory, ideology and practice

Conventional economic theory did not foresee any need to theorize about poverty alleviation. According to conventional theory, both rich and poor were faced with the same problem: the problem of scarcity. Scarcity is a psychological condition which arises because the individual economic agent tries to maximize his gains in the process of making choices to satisfy unlimited wants by using limited means. Put this way, both rich and poor are engaged in finding ways to deal with scarcity. Given this definition of the economic problem, there seems to be no need to have a separate theory of poverty. The economic mindset may therefore find a satisfactory solution of poverty in the framework of a 'free market'. But the free market has not been able to solve the problem of poverty. Critics of the free market have asserted that the economic mindset cannot solve the problem of poverty. As Gandhi one said, the world has enough resources to meet the needs of all the human beings but not the greed of one individual. However, need-based economies were part of the pre-capitalist societies. In contemporary market economies, market failure calls for a theoretical explanation and practical solution for mass poverty. State-led spending based on official development assistance (ODA) is one such solution.

Importance of state spending and market correction dawned on mainstream economists during the Great Depression in the second quarter of the 20th century. According to conventional economic wisdom, market mechanism was supposed to guarantee full employment and fair wages for work. But it was not happening in the real world. It was in the first half of the 20th century that economists realized that markets cannot automatically guarantee full employment. Wages are downward sticky, and declining profits and rising wages lead to market equilibrium below full employment. Market failure could be corrected by public policy to invest in public works to create income, employment and a high level of economic activity. In Pakistan, public policy visualized poverty alleviation through market-based economic development that had to be financed through creation of economic inequalities.[3] Addressing the question of economic inequalities, the authors of the first Five-Year Plan summed up the government's position as following "Inequalities of income originating in the inequalities of talent and service are desirable within 'limits'. Inequalities of income and wealth in the commercial and industrial sector play a useful role in the development of the country" (First Five-Year Plan, pp. 3–4). Pakistan's chief economist, Dr. Mahbub ul Haq, very forcefully developed and upheld this policy as a cornerstone for economic development during Ayub Khan's government (Mahbub ul Haq: 1966: 1–5).

This policy bias tells us that free working of the market was not conceived of as providing equal opportunities to economics agents or leading to equitable distribution of income. On the contrary, the façade of the "free market" was to be used to promote inequalities. Economic policies were to be designed and carried out in such a way that they led to concentration of economic resources in a few hands and brought poverty to the majority of the population. These inequalities were meant to appear as the natural result of the working of the invisible hand of the market rather than the visible hand of the policy makers.

Authors of the first Five-Year Plan were cognizant of the importance of human and social capital formation for poverty alleviation but did not feel obliged to do anything in this regard. It is interesting to note that according to the authors of the first Five-Year Plan,

> Cooperatives provide a means for achieving the benefits that accrue from large-scale organization, without the use of compulsion, force or major state intervention. In addition to economies for large-scale organization of some activities, cooperatives offer an opportunity for economic gain to cultivators by means of an organization in which they can perform many services for themselves in place of paying others to perform those services for them.
>
> (First Five-Year Plan: p. 31)

However, in former East Pakistan (current Bangladesh) the policy makers took initiatives to build social capital at the grassroots to alleviate poverty. These efforts had a mixed outcome. A review of success and failure of these donor-assisted poverty alleviation programmes provides us valuable insights about aid effectiveness for poverty alleviation.

Institutional parameters of aid effectiveness

The role of receiving mechanism in ensuring aid effectiveness

Pakistan Academy for Rural Development (PARD) Comilla (former East Pakistan, present Bangladesh) started a trailblazing poverty alleviation programme under a grant provided by the Ford Foundation for the Village AID programme.

The vision and leadership of Dr. Akhter Hameed Khan, founding director of the academy, played a key role in making this programme an unprecedented success. Dr. Khan was aware of the need to build the programme on ground reality. The first step in this direction was to train the civil servants to be development administrators. This meant teaching them to be public servants and to unlearn the ways of the colonial masters; training them to identify local needs based on consultation with the people; visiting villages and talking to farmers, the landless and women; and respecting the views of the people and knowing their strengths. The second step was to get the provincial government's consent to devolve all the line departments to the thana (police station) level, since officers based in every thana could easily reach out to all the villages in the thana's jurisdiction. These devolved departments were coordinated through Thana Training and Development Centres (TTDCs). This created the institutional and leadership base for the rural development programme. In Dr. Khan's words, this created the 'receiving mechanism' for development assistance. TTDCs did not have the capacity to reach out to every household in the village – an important prerequisite for poverty alleviation and an important segment of the receiving mechanism. For this purpose, the academy set out to create farmer's cooperatives and women organizations in each TTDC. This approach showed solid results and during the first 12 years (1959–1971) TTDCs were formed in 50 percent of thanas of East Pakistan. These TTDCs effectively lifted people out of poverty, and this effort was internationally acknowledged. In West Pakistan (present Pakistan), the Village AID programme fizzled out without showing any impact. Provision of development assistance without a sound receiving mechanism ends up in waste and leakages (SKAA: 2000, GOS: 2005). If an appropriate institution does not exist, then it is important to identify a pioneer within the existing institution to steer the process of institutional development. Without such a pioneer, an appropriate receiving mechanism cannot be developed.

Identifying the pioneer: Aga Khan Rural Support Programme (AKRSP)

Identifying the pioneer or early adopter is of critical significance at all levels of poverty alleviation work including community, government, support organization and government (Baqir: 2013). The Aga Khan Rural Support Programme (AKRSP) is a project of the Aga Khan Development Network (AKDN), the world's largest civil society network for poverty alleviation and human development. AKDN is a philanthropic institution set up by Karim Aga Khan to use philanthropy as an investment in human development (Khan: 2008). AKDN's philosophy is to test and start its programmes in view of local needs with its own resources. Their programmes are not donor driven. Once the programme has been successfully tested, they mobilize resources for its expansion. They believe in attracting the best talent to lead their projects and pay them well. In 1982 AKRSP was launched in the Gilgit district in the northern areas of Pakistan for poverty alleviation of rural households in one of the world's highest mountain ranges. Mr. Shoaib Sultan Khan (SSK), general manager of the Aga Khan Rural Support Programme (AKRSP), pioneered the community mobilization approach for this programme. The selection of SSK played a critical role in the programme's success and led to the doubling of the average income of target beneficiaries in the programme area.

The main tool of AKRSP's approach was the diagnostic dialogue. The diagnostic survey started with a visit by the management group to a village whose residents agreed to meet with AKRSP staff. Villagers were told that AKRSP would provide a grant for the project identified by the community on the condition that they agreed to (1) form a village organization (VO); (2) elect their office bearers by consensus; and (3) meet and save regularly. Communities were asked to choose leaders who could be easily brought under check by them in case of disagreement with the community. AKRSP provided skill training to the members of VO as well. This way AKRSP's funding served the purpose of building both the physical and social capital. The programme subsequently expanded to the neighbouring districts of Baltistan and Chitral and succeeded in doubling the income of 100,000 families in the next 10 years. It was evaluated as the world's best rural support programme in the external evaluation carried out by the World Bank.

The programme's secret of lifting people out of poverty consisted in starting small, identifying the activists who could glue rural communities together, showing impact and expanding through demonstration and setting up a new programme in new areas to avoid remote planning and development. Shoaib Sultan Khan's mentor Akhter Hameed Khan used to say, "you don't replicate programme, you replicate people". AKRSP's success led to establishment of similar rural support programmes (RSPs) in each province of Pakistan and outreach of these RSPs extended to almost all the districts. Impressed by the performance of this programme, SSK was asked by UNDP in 1996 to replicate the programme in the entire South Asian region. In India he found Kapula Raju for replicating his programme. Accordingly Mr. Khan Raju took the rural development work to new heights in India. The programme for poverty alleviation focused on women living in poverty in the state of Andhra Pradesh. Prior to SSK's visit, numerous programmes to alleviate poverty were started. These programmes included land reforms, welfare hostels for the children of poor to receive education and many others. All the government programmes put together did not result in substantial poverty reduction. The poor always looked to the government until SAPAP was introduced in 1996.

Selecting the right scale and progressing incrementally

When SSK presented his programme to government officials in Andhra Pradesh, his Indian hosts thought it was rhetoric. When the Government of Andhra Pradesh selected a few hundred villages for programme implementation, SSK said he would show how to harness the potential of the poor in one village only. The chief secretary of Andhra Pradesh, Kapula Raju, asked him to clearly state the conditions for success of his programme. SSK said the programme would succeed if only two conditions are met: (1) willingness of the poor to improve their lives and (2) presence of a local activist. In Raju's words, it was as simple and clear as Newton's law of gravity, but he doubted that it will move forward. He could not imagine that the programme would go to every village of Andhra Pradesh in the course of the next 10 years. In next 10 years 900,000 self-help groups (SHGs) were formed in Andhra Pradesh and millions of families were lifted above the poverty line. The programme trained thousands of community resource persons (CRPs) in bookkeeping, internal lending, monitoring and recovery of loans.

SHGs hold regular meetings, approve micro business plans and deal directly with the banks. Each business plan is prepared by the poor families themselves. In 2012 SHGs drew US$2 billion for micro credit. This portfolio increased sevenfold during the course of the last seven years. SHG members get this loan on a zero percent interest rate. This institutional infrastructure helped in making effectives use of the government's own poverty alleviation funds. The Government of India (GoI) had introduced a programme for death and accident insurance and old age pension for the poor. These programmes could not be universalized and claim settlement was also an issue. After the creation of the SHGs, 10 million women got access to insurance. The process for claiming insurance begins with a phone call to the call centre. A receptionist at the call centre locates a Beema Mitra (Friend of Insurance Programme) close to the village of the claimant. The Beema Mitra has a cell phone and an ATM card. As soon as she receives the report of a death, she draws Rs. 5,000 from an ATM and makes an immediate payment to the family of the deceased for burial/cremation and collects the death certificate from the village. Afterwards she delivers the death certificate to her office and receives her service fee. Within a week the balance of the insurance claim (Rs. 45,000) is paid to the family. All the people in this chain are women. Insurance scheme started when on the suggestion of Society for Elimination of Poverty (SERP) – an NGO registered by the Government of Andhra Pradesh – 10 million women agreed to pay Rs. 365 as annual premium for the insurance. The chief minister of Andhra Pradesh agreed to deposit an equal amount in the insurance fund on behalf of the government. These two contributions plus the interest accumulated to a handsome amount at retirement age.

In India, 40 percent of children are under-nourished, mostly because women are also malnourished. Women in Andhra Pradesh came up with the idea that good nutrition should start with pregnancy. It should continue for 1,000 days of the lactation period. Under this programme women in 5,000 villages are provided good, nutritious meals three times a day at the nutrition centre. It is financed by Village Samakhia and paid back in small instalments spread over a long period. Ten thousand milk collection and processing points have also been established with the help of women. Women have taken over the whole cycle of milk production and delivery to chilling centres.

All the new initiatives are tested in one or two villages and then spread through CRPs. Women who experiment and benefit from the pilot project and are able to spare time are picked as CRPs, and they are paid for their services. There are thousands of such CRPs in Andhra Pradesh. They have learnt Hindi for the first time in their life because they are in high demand in northern states. To implement this programme on a large scale, the Union Government of India designed the National Rural Livelihood Programme to be implemented by the National Rural Development Mission.

Aid is effective if it can support the process: the case of Orangi Pilot Project (OPP)

Dr. Akhtar Hameed Khan initiated an urban poverty alleviation programme in Orangi, one of the largest informal settlements of Karachi, in 1980. As director of the Orangi Pilot Project (OPP), one of his outstanding achievements was mobilizing the residents of Orangi to provide sanitation coverage to the entire settlement with their own financial contributions, supervision and management. OPP was started with the philanthropic contribution of the Infaq Foundation. Khan agreed to work for poverty alleviation in Orangi on three conditions. First, he insisted on managerial independence – it meant that he would make all the programme decisions without the intervention of the sponsor but would regularly report on programme's progress and financial transaction in a transparent manner. Second, he insisted on financial independence – Infaq would provide funds for investment in an endowment to generate a stream of income to cover programme expense to ensure programme continuity and free it from the burden of continuously looking for funding, or seeking funds as contractor, or compromising the process-based approach to achieve results in a 'time-bound', template-based approach. Third, he insisted on programmatic independence – he would seek guidance from the poor on how to alleviate poverty, not follow a predetermined solution proposed by the donor (Khan: 1992).

In the early stages of project conceptualization, United Nations Centre for Human Settlements (UNCHS) offered the services of an international consultant, Nicholas Houghton, to provide technical support to OPP. Very early in the project, serious differences emerged between Mr. Houghton's 'results-based approach' and Dr. Khan's 'organic-pragmatic-sociological approach'. It was a very clear demonstration of a conflict between a blueprint and a process-based approach. In his note to sponsor of the OPP, Agha Hasan Abedi in 1983, Mr. Houghton mentioned that "Orangi Pilot Project is a highly personal research and extension initiative into low-cost urban technology and social services for low-income communities. It is subject totally to the perceptions, intellectual speculations, and will of its originator and director Dr. Akhtar Hameed Khan, a very able man with a well-established reputation as' a dedicated social educator."[4] "The project only works with those groups that respond to the project. Thus, there can be no fixed target areas. In the end, through this process, the project hopes to discover a harmonious whole of need, viability and self-propagated replicability. The right information and technical improvements will be those that are incorporated into the shared culture of the community. This assimilation, it is assumed, will lead to further development and self-reliance."[5]

In response Dr. Khan stated that here was no blueprint, no master plan to be imposed; OPP is an NGO, not equipped or designed to match the planning, regulating or servicing functions of official agencies like the KDA, KMC or the departments of education and health. An NGO could not be a parallel agency, as the Governor of Sindh rightly warned in the presentation meeting in February 1981; that OPP's main concern is to promote self-supporting people's organization, and that OPP's research was designed to discover technological, sociological and economic models which were based on popular participation, management and funding etc. (Khan: 1998: 163).

During the first six years, sewerage lines in 4,000 lanes were laid with the technical assistance and social guidance of OPP. Eventually the programme expanded to 7,000 lanes in Orangi and benefited 70,000 households. By 2007, the network of OPP's partners expanded to 331 cities, towns, settlements and villages outside Orangi with an investment of Rs. 100.26 million by the community members. The National Sanitation Policy approved by the government in September 2006 included the OPP's model for adoption. Mr. Houghton discontinued his programme after laying sewerage lines in 36 lanes. This example clearly demonstrates that development aid tied to elaborate frameworks and templates might appear extremely impressive on paper but does not succeed in delivering the intended targets. In the end, the blame for failure is attributed to 'lack of awareness' of the people without realizing the need for enhanced 'awareness of the expert'.

Development without loans

If you manage well, you don't need to borrow

*Sindh Katchi Abadi Authority's (SKAA) programme
for the shelter for poor*

In 1977, the Government of Pakistan established the Sindh Katchi Abadi Authority (SKAA) with a view to start regularization and upgradation of urban squatter settlements and coordinating the process of issuing leases to the dwellers.[6] Nothing happened till 1993 because no government official was willing to stick his neck out to give ownership deeds to the squatters. Nothing without precedence is considered worthy of putting your head on the chopping board in official conformist civil services culture of Pakistan. When Tasneem Siddiqui took charge as director general of the SKAA in 1993, things changed. Siddiqi was not very fond of treading the beaten path. That is why he was suspended from his job a couple of times. He was a keen observer and open to learning from people in the areas that he served. During his tenure as director general of SKAA, he discovered that development and delivery method adopted by government agencies had some basic flaws. The government's price and payment schedule was not in line with the financial conditions of the poor; therefore, low-income people could not purchase the plots in katchi abadis (KAs). There was also no time limit to complete construction. Any financially well-off person could buy the plot in the name of the poor and keep it vacant as long as he wanted. This arrangement offered good opportunities to speculators to invest in these plots while the KAs kept growing.

Siddiqui developed an alternative system based on his observation of the housing pattern followed by the land-grabbers in developing KAs. He designed following steps for regularizing and upgrading KAs: (1) to sell undeveloped land at affordable price in easy to pay instalments; (2) to create an easy entry system by doing away with the balloting; (3) not selling fully developed plots but undertaking incremental development so that the buyers could pay development expenses in easy instalments during the course of development work; and (4) giving possession to only those buyers who moved in with their family members to settle down after making the down payment. Tasnim Siddiqui revised standard operating procedures (SOPs) of SKAA on these lines, reduced drastically the number of steps needed for making lease application, moved SKAA field offices to KAs and regularized almost 50 percent of KAs in Sindh during his tenure. SKAA met its overhead costs by recovering the lease charges and by June 1994 became financially self-sufficient. A World Bank loan of US$5 million extended to SKAA was never used. Tasneem Siddiqui conclusively showed that with indigenous knowledge and good management, a lot can be accomplished without a loan. It was in contrast to KWSB's effort to solve its financial woes by asking for loans rather than improving its knowledge of ground reality and management practices. When an NGO coalition opposing agreement to seek a loan from ADB presented their data and documentation to the governor of Sindh, Moinuddin Haider, they were able to prove that the project costing $100 million could actually be completed with $30 million and got the loan cancelled.

Don't underestimate the giving spirit of your people

Poverty alleviation through interest free credit – Akhuwat

The history of microfinance in Pakistan goes back to the cooperative movement during British rule and the Village AID programme in 1960s. Various innovations in the microfinance sector emerged during the 1980s and 1990s. Sustainability of microfinance services is a major challenge to micro credit organizations. Microfinance institutions helped poverty alleviation by creating income generation activities for millions of micro entrepreneurs, provided access to credit for people who had no collateral to offer to secure loans and offered credit on interest rate at one tenth of the interest rate charged by the money lenders. But even the interest rate of 18 percent per annum pinched Dr. Amjad Saqib, who considered it unfair to charge any interest to the poor. Dr. Akhter Hameed Khan's argument for charging interest was that no interest charges will tempt the poor to deposit money in the banks and earn interest without working. Dr. Saqib decided to challenge this concept. Ideally it was the responsibility of an Islamic welfare state to provide this facility. But in the absence of a functioning welfare state, he thought it was the social responsibility of the well-off groups to help the deprived sections of society. He decided to start his interest-free micro credit programme without borrowing a single penny from a lending institution or requesting grant money. He started his programme with a first loan of Rs. 10,000 in a low-income neighbourhood in Lahore. The loan money was donated by a housewife. Others joined and subsequently Dr. Saqib was able to raise US$2 million from the Pakistani diaspora.

The programme has a loan recovery rate of 98 percent. It has a very low overhead cost, is supported by volunteers and is expanding to other districts of Pakistan. Akhuwat has encouraged its supporters to establish and manage programmes in their districts as their own local programmes, not as franchises of Akhuwat. Dr. Amjad Saqib said goodbye to his job as a civil servant and now works for Akhuwat full-time. Akhuwat charges borrowers an administration fee of 5 percent of their loans, irrespective of the timing of the repayment. This administration fee, however, is exempted on loans of less than Rs. 4000. By June 2010 Akhuwat had lent Rs. 700 million to over 40,000 clients. Akhuwat deliberately avoids grants from official foreign donors or other similar sources. It might be considered the most innovative poverty alleviation programme across the globe (Baqir: 2009).

Knowledge-based management; the case of primary healthcare

Two primary healthcare programmes, led by renowned Pakistani social entrepreneurs Nasim Ashraf and Jahangir Khan Tareen, offered improved primary healthcare by improving healthcare management and changing existing practices in view of the knowledge gained through action research. These programmes are known as the Rahim Yar Khan pilot project of PRSP and Gujrat pilot projects of NCHD. Both the models were initially expanded to other districts in Punjab and then to other provinces and finally replicated under the federal government initiative. Both these models demonstrated that improvement of management, strict monitoring and community participation are key elements for improving primary healthcare services. It is not more spending but better spending that makes a significant difference. While most of the donor-funded programmes for capacity building focused on training paramedics, who kept migrating overseas in search of lucrative jobs due to irrational fiscal incentive, these models rationalized fiscal initiatives, improved management and turned around the Basic Health Units (BHUs). Micro health insurance, the latest initiative carried out by NRSP and Adamjee Insurance, provides health coverage for the whole family for Rs. 350 (US$4) and might seem too good to be true. That is the innovation which will not only affect our health policy but our entire way of thinking about dealing with poverty.

Restore, don't borrow

Punjab Housing, Urban Development and Public Health Engineering Department (HUD&PHED) was established in 1961. The department is responsible for providing water supply schemes to rural areas. By 2010 the department had completed 4,058 rural water supply schemes, of which only 2,448 schemes were functional. By 2017 the number of dysfunctional schemes increased to 4,662

(Farooqi: 2017).[7] Millions of dollars of loan money were spent to build these schemes. Most of these schemes became dysfunctional due to lack of community engagement in financing, building, operating and maintaining the schemes. Usually when the issue of restoring the schemes came up, the lazy bureaucratic response was to ask for a new loan to build a new scheme. In one of the instances, Lodhran pilot project's (LPP) senior social organizer Hafiz Arain was told by the Tehsil Municipal Administration (TMA) Khanpur that their sanitation scheme, built at the cost of Rs. 15 million, was completely dysfunctional and they needed almost an equal amount to put in place a new scheme. Hafiz set to work to figure out where the sanitation lines were blocked and how the draining of liquid waste in an open area could be stopped. In a few months he restored the entire scheme at the expense of Rs. 150,000 without asking for any loan. Years later, Nazir Ahmad Wattoo repeated Hafiz Arain's approach in restoring TMA Bhalwal's water supply scheme for a town of 100,000 people at *no cost*. These two schemes convincingly show that restoring dysfunctional schemes is another important source of reducing the debt burden (Iftikhar et al.: 2018).

Universal education; social entrepreneurship as an alternative to borrowing

A number of promising approaches to education emerged in 1990s including that of community-based schools. A variety of models with a low-cost budget, increased community participation and focus on low-income and working children and adults are functioning at present. These models offer the possibility of financial sustainability and replication on a large scale. Home schools for children in low-income areas and male and female adult literacy centres in rural areas were set up by an Adult Basic Education Society (ABES) in numerous districts. An alternative approach was to build one room, one-teacher schools with a view to turn them into formal schools through incremental development and make them a part of the mainstream. This approach was followed by the Society for Community Support for Primary Education (SCSPEB). The SCSPEB promoted government-community partnerships by establishing female schools with community participation. Under this approach community would donate land, agree to send girls to the school and initially one teacher would be appointed to start teaching. After regular functioning of school for a specific period it is handed over to the government. Both these models offered alternatives for better management of government schools (Baqir: 2004).

However, Care Foundation and the Citizen Foundation have blazed the trail for providing high-quality education through well-funded schools by the social entrepreneurs. Quality of education provided by both these models provides education comparable to prestigious private sector schools. Various provincial governments have also outsourced management of government schools to them. This again highlights the point that it is better spending through private-public partnership rather than borrowing in the name of the poor, which makes the difference.

Edhi Foundation's alternative to external borrowing: volunteerism and philanthropy

Edhi Foundation is the largest and most organized social welfare network in Pakistan. The Foundation works round the clock and provides services without any discrimination on the basis of colour, race, language, religion or politics. The dynamic nature and the range of social services provided by Edhi Foundation makes it different from other similar organizations. The foundation's activities include a 24-hour emergency service across the country through 250 Edhi Centres which provide free shrouding and burial of unclaimed dead bodies and shelter for the destitute, orphans and handicapped persons. It has set up free hospitals and dispensaries, rehabilitation facilities for drug addicts, free wheelchairs, crutches and other services for the handicapped, family planning counselling and maternity services and national and international relief efforts for the victims of natural calamities. Edhi strongly adheres to the principle of self-help. It is hard to think of any welfare organization or an NGO sustaining its functions while refraining from soliciting financial support from the government and foreign aid–giving agencies. All donations to the Edhi Foundation come from individuals and a few business enterprises. Edhi proudly stated that the poor are among the most giving people in Pakistan. He practiced a simple principle throughout his life: if you have two pennies, spend one penny on yourself and the second one on the needy (Raponi and Zanzucchi: 2013).

Anatomy of aid failure

Dr. Mahbub ul Haq became disillusioned with his own strategy of economic development based on generation of income inequalities. He was the first one to point out concentration of economic power in the hands of 20 families and the emergence of the black economy as 50 percent of Pakistan's economy. This led him to the conclusion that human development is the right path to development. Dr. Mahbub ul Haq's idea of human development as a meaningful measure of economic development got wide acceptance among the donor community across the globe. It was picked by the United Nations Development Programme (UNDP) as a central theme for its development assistance to low-income and middle-income economies. He therefore presented the idea of human development through development of social capital in Pakistan to alleviate poverty and provide solid foundations for economic development. This idea was converted into a development plan with the support of key multilateral and bilateral donors in the form of a Social Action Programme (SAP).

SAP was implemented during Zia ul Haq's military regime and aimed at building micro development infrastructure, providing education, health and water supply services and generating income opportunities for low-income communities through community mobilization. The vision behind this programme was to develop the human condition through enlarging the range of choices of low-income communities. Community mobilization was to be used for creating economies of scale for low-income communities and enabling them to make choices which could not be availed without organized efforts. This view had been tested and validated by many government- and NGO-led experiments in Pakistan and abroad. Community participation had succeeded where private and public sector had failed. The power of this idea helped Dr. Haq easily mobilize resources for its implementation.

However, there was one fundamental flaw in SAP's strategy. Since it was a donor-driven programme, it was meant to produce results in a narrow time frame against a list of indicators. This put pressure on the programme managers to rapidly spend money to show 'results'. This pursuit of results necessitated bypassing the process of community engagement in various cycles of the project, and engagement of contractors to speed up delivery and in many cases making advance payments to the contractors to meet delivery targets. This top-down, target-based, time-bound approach did not succeed. It reaffirmed the view that in poverty alleviation ends cannot justify the means. Means are as important as much as the goals. Without appropriate institutions and flexible approach money alone cannot produce miracles. That is what happened with SAP.

Evaluation of first phase of SAP revealed that the programme did not achieve its desired objectives. Dr. Mahbub ul Haq failed once again to lift the poor people of Pakistan above the poverty line. While the diagnosis of problem was right there seemed to be something lacking in Dr. Haq's approach for achieving his intended results. As a national planner trained in solving macro-economic problems, he depended on macro-economic tools for micro-economic changes. This bungled the whole programme. Development planners aiming at large scale changes have conventionally depended on a top-down template-based approach for human and social development. Such an approach does not want to be bogged down in the unpredictable and uncontrolled process of engaging the communities and hesitates to give them lead role for their own development. Engaging communities seems risky, inefficient and input based to conventional economic thinkers. However, the appearance and expected results of template and process-based approach are very deceptive. The template-based approach appears predictable, efficient and result oriented but runs out of steam in very early stages of community development. Process-based approach is opposite in character and outcome.

This view was reinforced in the assessment of first phase of the Social Action Programme (1993–1998). A World Bank report, "Poverty in Pakistan, Vulnerabilities, Social Gaps, and Rural Dynamics", stated that the relative insulation of social spending from downward pressures during 1993–1998 was largely due to an infusion of $2 billion in support of the Social Action Programme (SAP). The report regretted that there was a serious problem of governance in Pakistan. Resources that were allocated to social spending over the past decade were largely used inefficiently and failed to have a significant impact on a dollar per dollar basis. Pakistan in fact exhibited persistent problems in most dimensions of governance that are relevant for sound public spending. The report added that there were leakages, difficulties with bureaucratic structure and quality, weaknesses in the rule of law, and opacity in government decision-making.[8] This was mainly due to undeveloped social infrastructure and neglect of process for developing social capital.

Donor reluctance to work with self-reliant NGOs

Experience of donor-assisted programmes has clearly shown that aid can produce results if it is channelled through appropriate receiving mechanism and if it provides ample room for the process of community engagement to unfold. Many NGOs and government departments in Pakistan have established their credentials by meeting these conditions. It has made them effective and freed them from the need to seek donor funding. Their experience provides valuable insights into poverty alleviation in Pakistan. These institutions include but are not limited to AKRSP, Edhi Foundation, Akhuwat, SKAA, Care Foundation, TCF, SIUT and OPP. However, donors and development agencies do not engage with such partners because they cannot control them. A report by a panel of eminent persons appointed by the secretary-general of the United Nations pointed out that "Public Opinion has emerged as the second super power in the world" (UN: 2004). Subsequently an experiment for programming as One UN started in Pakistan and a Civil Society Advisory Committee was constituted to advise UN Resident Coordinator and 19 UN agencies. UN agencies showed hesitation in signing partnership agreement with NGOs not receiving grant funding from them because there was no room in their procedures to form such a partnership. Here is the conundrum of aid effectiveness; giving aid to contractors neither aligns aid with local priorities nor does it ensure sustainability after the end of grant funding but partnering with institutions that can show the way to take care of both these concerns is not possible under the development agencies' practices.

State, civil society and donors

Akhter Hameed Khan used to say that NGOs are research and development (R&D) department of poverty alleviation and government is its marketing department. NGOs should experiment, refine and develop the model at local level and government should take it to scale. This view provided the foundation for NGO-government partnership. He also said that communities are better endowed with land and labour and NGOs/donors are better endowed with organization and capital. Poverty alleviation models should seek community and development agency's (government, donor, NGO) partnership based on this factor endowment (WSP: 2007). It should not be a handout. Such models will be sustainable (Baqir: 2013). Poor cannot access services provided by the formal sector (both in public and private domain) because their technical specifications are overdeveloped and consequently their costs are not affordable for the poor. Poor are not destitute. They are willing to pay for the services if development agencies can design frugal and affordable models. It takes a long time to test and develop such models, acquire expertise and gain community's trust in running programmes based on these models. Donors or lending agencies do not have patience to engage in such activities. Since many lending agencies depend on extremely expensive international 'experts', their administrative cost for running such pilot projects tremendously increases the overhead cost of such programmes and makes poverty alleviation so expensive that it cannot be carried out without foreign grants and loans. Such expensive development projects result in abandoned scheme and exorbitant loan burden. Instead of alleviating poverty formal sector models end up adding to the burden of poverty. All the loans received for 'poverty alleviation' have to be paid through the tax earnings. In a country like Pakistan where a regressive taxation system depending on 60 percent indirect tax revenues is in place, the poor pay through the nose for the follies of poverty experts. These facts call for redefining the terms of partnership for poverty alleviation.

Under the new terms of partnership model building should be done by local government agencies, NGOs or social entrepreneurs. Sustainability should be assessed by comparing the cost and revenue streams of the projects and funding should be provided to expand this work for greater public benefit. Local model development is time-consuming, labour intensive, small in scale, unpredictable, best done locally, process based not template based but, in the end, it is low cost and affordable. By design it has to be transparent, inclusive, downward accountable and a reciprocity-based partnership. Such work is possible through endowment-based cash flow. These endowments can either be created by indigenous philanthropists or government. Donor and lender systems cannot commit funds for such investments. Therefore donors/lenders can make best use of their resources by finding credible local partners, frugal models for poverty alleviations and only investing in taking success to scale. Frugal sciences and indigenous philanthropists can put in place the models which create access for the poor by expanding the range of choices (Sen: 2001), building human capital and protecting against unequal global competition on the one hand and privatization, deregulation and demolition of barriers to trade on the other (Rapley: 2007). A cookie cutter approach cannot work (Rapley: 2007: 229).

Civil society's role in finding innovative and frugal ways for poverty alleviation is upheld by policy makers and practitioners subscribing to diverse set of ideologies; neo-liberal, reformist and anti-globalization (Patel et al.: 2001). Pakistan's successive governments were aware of this potential and did try to engage civil society in the process of development. These efforts were partially successful. A review of Pakistan's past 60 years of social change shows that negation of entrepreneurial vision during the implementation of inclusive initiatives led to the failure of attempts for inclusive change in Pakistan. A review of multi-donor supported Social Action Programme (SAP) by SPDC noted that "Expenditures on the social sector over the four-year period (1993/4 to 1996/7) aggregated to over Rs. 163 billion. Starting from a level of Rs. 27.7 billion in 1993/4 they have grown rapidly at the rate of 25 percent per annum". Use of conventionally trained professionals such as engineers and architects for field implementation and interaction with communities is not advisable unless they subscribe to and/or have been initiated into a development approach which views communities as teachers and partners rather than passive recipients of funds, ideas and technologies.

Conclusion

Success of poverty alleviation and inclusive growth depends on creation of a new partnership between external development agencies, civil society organizations and communities. For effective policy implementation development experts need to join hands with people and politicians at every stage of project cycle. Nature of this partnership is summarized in the table given below. Only a process encompassing these roles and responsibilities may lead to effective implementation of inclusive development policies. External aid can be effective by strengthening the process of social capital formation not by bypassing it. Clarity of roles and expectations from various partners is the key to aid effectiveness and in many cases development without external assistance. It is also important to point out that development assistance should be considered a means not an end of poverty alleviation. Terms of loan management should not dictate the path to poverty alleviation it should be the other way around. There are no shortcuts. The key elements of appropriate role based partnership between communities, experts and resource providers are summarized in Table 7.1.

The process for strengthening collaboration between the people, professionals and development agencies

Table 7.1 Appropriate process and role for strengthening collaboration between the people, professionals and development agencies

Steps/Tasks	Professionals	People	Development Agencies and Politicians
Creating Local Institution	Setting up an apex body for developing social and technical solutions	Expressing local priorities Serving as receiving mechanism	Allocation of resources Establishing people friendly regulatory framework
Documentation	Mapping Data gathering and surveys Preparing plans	Contributing knowledge Learning documentation skills	Providing access to information Supporting local knowledge development
Costing and designing	Low-cost technical specifications Department work	Employing the skills of community specialists	Allocation of resources

Steps/Tasks	Professionals	People	Development Agencies and Politicians
SOPs	Client-oriented simple rules of business Department work	Forming organization Regular saving Regular meetings	Notifying revised SOPs
Financing	Developing low-cost options	Raising funds through component sharing	End to patronage and placement of process-based funding
Implementation	Technical supervision Eliminating contract work	Hiring locals Keeping muster rolls, accounts Record keeping	Delegating implementing tasks to COs
Monitoring and reporting	Developing indicators with local consultation	Reporting on success indicators	Participatory performance assessment

Notes

1　See for example, Harlan Cleveland, *The Knowledge Executive: Leadership in an Information Society*, Truman Talley Books, New York, 1985. Cleveland has emphasized the point that development cannot be exported. So in the case of economic development of developing economies and poverty alleviation 'supply does not create its own demand'. It depicts a clear case of failure of 'supply-side economics'.

2　Samia Waheed Altaf has a given a fascinating account of one such hurriedly commissioned mission and its dynamics in her book *So Much Aid: So Little Development*. It shows in a very clear way how the lofty ideals of development assistance are gradually compromised to accommodate pressing requirements arising out of the compulsion of consultancy work.

3　See Haq (1966).

4　See Annex of Akhter Hameed Khan (1996). *Orangi Pilot Project, Reminiscences and Reflections*, Oxford University Press, Karachi.

5　Ibid.

6　*Katchi abadi* is an Urdu phrase which literally means temporary settlement, denoting a settlement whose residents do not have permanent legal right of ownership over the land they occupy. It is the Urdu alternative of the term informal settlement.

7　According to Monem Farooqi, there were 4,662 completely dysfunctional drinking water supply schemes by 2017. Earlier data posted on PCRWR website giving breakdown of various functional and non-functional schemes and source of loans from these schemes has been erased from PCRWR website. See Monem Farooqi (2017), 'Punjab government mute over averting serious drinking water issues', November 27, 2017, https://profit.pakistantoday.com.pk/2017/11/27/punjab-government-mute-over-averting-serious-drinking-water-issues/. Subsequently the Government of Punjab approved a staggering amount of Rs 9.3 billion for rehabilitation of 781 dysfunctional rural water supply schemes (*Business Recorder*, October 23, 2017, https://fp.brecorder.com/2017/10/20171023228782/).

8　For details see World Bank (2002).

References

Aga Khan Foundation (AKF), 2003, *Understanding Civil-Society-Government Relations in Pakistan: Public Private Partnership for Sustainable Development*, Proceedings of a Workshop, August 27–28, 2003, Karachi.

Ahmad, M., 2008, "State of the Union: Social Audit of Governance in Union Council Bhangali", *Lahore Journal of Policy Studies*, Vol. 2, No. 1, Lahore.

Baqir, F., 2009, *Irada-A Collection of Case Studies on People Centered Development* (in Urdu), Suchet Kitab Ghar, Lahore.

Baqir, F., 2013, *A Module on Participatory Development*, AHKRC, 40, Islamabad.

Baqir, F., 2004, *Myths of Primary Education* in *Development Issues: Innovations and Successes*, T.A. Siddiqui (ed.), City Press, UNDP Joint Publication, Karachi.

EDC, 1996, *Draft Report-Government/Donor/NGO Collaboration: Lessons Learnt and the Action for Future*, United Nations Development Programme, Islamabad.

Farooqi, Monem, 2017, "Punjab Government Mute Over Averting Serious Drinking Water Issues", *Pakistan Today*, November 27 viewed from https://profit.pakistantoday.com.pk/2017/11/27/punjab-government-mute-over-averting-serious-drinking-water-issues/

Government of Pakistan, 2002–2007, *Public Sector Development Programme* (2002/2003–2005/2006) Planning Commission, Islamabad.

Government of Pakistan (GOP), 2005, *Medium Term Development Framework (2005–2010) Supplement: Rural Poverty Reduction through Social Mobilization (2005–2010)*, Planning Commission, Islamabad.

Government of Sindh (GOS), 2005, *District Planning Manual*, Planning and Development Department, Karachi.

Habib, Mehjabeen Abidi, 2002, *Green Pioneers UNDP-GEF*, City Press, Karachi.

Haq, M., 1966, *The Strategy of Economic Planning: A Case Study of Pakistan*, Oxford University Press, Pakistan.

Hasan, A., 1997, *Working with the Government: The Story of OPP's Collaboration with State Agencies for Replicating Its Low Cost Sanitation Programme*, City Press, Karachi.

Hasan, A., 2007, *Improving Urban Water and Sanitation Provision Globally, through Information and Action Driven Locally*, unpublished paper submitted to DFID.

Husain, T., 1992, *Community Participation: The First Principle: A Pakistan National Conservation Strategy Paper 1*, Environment and Urban Affairs Division Government of Pakistan, IUCN.

Iftikhar, M. N., S. Ali and A. Sarzynski, 2018, "Community–Government Partnership for Metered Clean Drinking Water: A Case Study of Bhalwal, Pakistan", in S. Hughes, E. Chu and S. Mason (eds.), *Climate Change in Cities: The Urban Book Series*. Springer, Cham.

Iqbal, Muhammad Asif, Hina Khan and Surkhab Javed, 2004, *Non-profit Sector in Pakistan: Historical Background*, Social Policy and Development Centre, Aga Khan Foundation and Centre for Civil Society Studies, John Hopkins University.

Islam, N., 1981, *Foreign Trade and Economic Controls in Development: The Case of United Pakistan*, Yale University Press, New Haven, CT.

Kamal, S., 1996, *The NGO-Donor Axis: Suggested Code of Conduct for NGOs and Donors in Pakistan*, *A United Nations Studies Paper 2*, prepared for UNDP and Local Dialogue Group, Islamabad.

Khan, A. H., (1998) *Orangi Pilot Projects Programme*, OPP-RTI, Second Edition, Karachi.

Khan, Akhter Hamid, (1998) *Orangi Pilot Project, Reminiscences and Reflections*, Oxford University Press, Karachi.

Khan, Aga, 2008, *Where Hope Takes Root*, Douglas and McIntyre, Vancouver.

Khan, S. S., 1980, *Rural Development in Pakistan*, Vikas Publicizing House Ltd, Ghaziabad.

Leitner, G. W., 1882, *History of Indigenous Education in the Punjab since Annexation and in 1882*, Reprint Delhi: Amar Prakashan, Calcutta.

Lewis, Stephen R., Jr., 1969, *Economic Policy and Industrial Growth in Pakistan*, George Allen and Unwin Ltd., London.

Lewis, Stephen R., Jr., 1970, *Pakistan: Industrialization and Trade Policies*, Oxford University Press, London and New York.

Naved, H. and I. Nabi (ed.), 1991, *The Aid Partnership in Pakistan in Transitions in Development: The Role of Aid and Commercial Flows*, International Centre for Economic Growth, San Francisco.

NRSP, 2000, *In Commemoration of the Life and Times of Akhter Hameed Khan: Talks of Akhter Hameed Khan at the National Rural Support Programme*.

Pakistan Centre for Philanthropy (PCP), 2007, *Analysis of MDGs and Civil Society of Pakistan*, unpublished.

Pakistan National Planning Board, 1958, *The First Five-Year Plan (=60)*, Karachi.

Pasha, Dr. Aisha Ghaus, Haroon Jamal and Muhammad Arif Iqbal, 2002, *Dimension of the Non-Profit Sector in Pakistan* (Preliminary Estimates) Social Policy Development Centre, Aga Khan Foundation (Pakistan) and Centre for Civil Society Studies, John Hopkin, University.

Patel, Sheela, Sundar Burra and Celine D'Cruz, 2001, "Slum/Shack Dwellers International (SDI) Foundations to Tree Tops", *Environment and Urbanization*, Vol. 13, No. 45, p. 1.

Rapley, J., 2007, *Understanding Development: Theory and Practice in the Third World*, Lynne Rienner Publisher, London.

Raponi, Lorenza Raponi and Michele Zanzucchi, 2013, *Half of Two Paisas: The Extraordinary Mission of Abdul Sattar Edhi and Bilquis Edhi*, Translated from Italian by Lorraine Buckley, Oxford University Press, Pakistan.

Sen, A., 2001, *Development as Freedom*, Oxford University Press, New York.

Siddiqui, Tasneem Ahmed (ed.), 2004, *Development Issues: Innovation and Success*. City Press, Karachi.

Sindh Katchi Abadis, 2000, *Authority Upgradation/ Improvement of Katchi Abadis – Department vs. Contractor's Work: A Comparative Study*.

SPDC, *Review of the Social Action Programme*, Karachi.

UN Report of the Panel of Eminent Persons on United Nations – Civil Society Relations, 2004, *We the Peoples: Civil Society, the United Nations and Global Governance*, United Nations, New York.

UNDP, 1991, *NGOs Working for Others: A Contribution to Human Development*, Vol. 1, United Nations Development Programme, Islamabad.

World Bank, 2002, *Poverty in Pakistan, Vulnerabilities, Social Gaps, and Rural Dynamics*, September, Islamabad.

WSP Field Note, 2007, *Empowering Citizens' Participation and Voice*.

Index

Note: **Boldface** page references indicate tables. *Italic* references indicated boxed text and figures.

Printed and bound by CPI Group (UK) Ltd, Croydon, CR0 4YY

24/10/2024

01778282-0019